Geocriticism and Spatial Literary Studies

Series Editor
Robert T. Tally Jr.
Texas State University
San Marcos, TX, USA

Geocriticism and Spatial Literary Studies is a new book series focusing on the dynamic relations among space, place, and literature. The spatial turn in the humanities and social sciences has occasioned an explosion of innovative, multidisciplinary scholarship in recent years, and geocriticism, broadly conceived, has been among the more promising developments in spatially oriented literary studies. Whether focused on literary geography, cartography, geopoetics, or the spatial humanities more generally, geocritical approaches enable readers to reflect upon the representation of space and place, both in imaginary universes and in those zones where fiction meets reality. Titles in the series include both monographs and collections of essays devoted to literary criticism, theory, and history, often in association with other arts and sciences. Drawing on diverse critical and theoretical traditions, books in the Geocriticism and Spatial Literary Studies series disclose, analyze, and explore the significance of space, place, and mapping in literature and in the world.

More information about this series at
http://www.palgrave.com/gp/series/15002

Julius Greve · Florian Zappe
Editors

Spaces and Fictions of the Weird and the Fantastic

Ecologies, Geographies, Oddities

Editors
Julius Greve
Carl von Ossietzky University
of Oldenburg
Oldenburg, Niedersachsen, Germany

Florian Zappe
Georg-August-Universität Göttingen
Göttingen, Niedersachsen, Germany

Geocriticism and Spatial Literary Studies
ISBN 978-3-030-28118-2   ISBN 978-3-030-28116-8   (eBook)
https://doi.org/10.1007/978-3-030-28116-8

Cover credit: Netfalls Remy Master/shutterstock.com

This Palgrave Macmillan imprint is published by the registered company Springer Nature
Switzerland AG
The registered company address is: Gewerbestrasse 11, 6330 Cham, Switzerland

FOREWORD

WEIRD GEOGRAPHIES, FANTASTIC MAPS

Maps seem so commonplace and universal that their users often forget just how very weird they really are. Maps are certainly representational, yet they are so wildly figural and metaphorical as to be utterly unrealistic, even as they are also associated with the most prosaic realisms of day-to-day existence. As a means of comprehending a given space, the map offers an entirely alternative epistemology in which the various features of that figured space are registered, omitted, highlighted, or suppressed, and nothing on the map itself in any way replicates the "real" spaces and the elements within them. Contrary to popular understanding, maps do not depict the actual places that appear on their surfaces, but rather they involve elaborately allegorical structures by which the reader (or map-gazer) may attempt to make meaningful sense of these spaces. Maps are, by definition, works of fantasy.

This is, after all, part of their power. By presenting a practical and meaningful image of a space that is nevertheless completely figurative, maps essentially entertain alternative realities by which to make sense of the real-world spaces and places. Speaking of literal maps in *A Thousand Plateaus* (1980), Gilles Deleuze and Félix Guattari famously noted that "[t]he map is open and connectable in all its dimensions; it is detachable, reversible, susceptible to constant modification. It can be torn, reversed, adapted to any kind of mounting, reworked by an individual, group, or social formation. It can be drawn on a wall, conceived as a work of art, constructed as a political action or as a mediation."[1] Such versatility accounts for the well nigh universal appeal of the map as a tool for orienting oneself in space, for navigating routes through space, and for representing territories at a helpfully abstract level—the "bird's-eye view," for example, beyond the subjective perception of any individual on the ground—not to mention the appeal of maps as aesthetic objects or works of art, partially or completely removed from the pragmatics of place and movement. In its wide variability of usage, popularity, influence, and activity, the map finds itself aligned well with other intrinsically multivalent social and artistic forms, such as literature, films, and similar media.

I have frequently endeavored to make connections between literature and mapping, combining the two in the term "literary cartography," but also extending the literal and figurative meanings of these ideas in such a way as to account for seemingly innumerable methods by which human beings make sense of or give form to their worlds and their situations within them. I have argued that literature, along with other media, can function as a means by which to map the *real-and-imagined spaces* (as Edward Soja famously named them) of our societies, physical environments, and conceptual domains, and that the subsequent maps become vehicles for achieving a sense of place in both space and time that in turn enables us to interpret, understand, and ultimately transform the worlds we inhabit and think.[2] This literary cartography is grounded in acts of the imagination, and the underlying genre or discursive mode could thus rightly be labeled *fantasy*. I maintain that even the most putatively realistic works of literature, like the seemingly realistic but (upon further thought) obviously figurative maps, are at their root essentially fantastic. The effectiveness of these imaginative endeavors may be measured in part by the widespread influence and popularity of literature and other forms expressly categorized as fantasy, broadly conceived so as to include such genres as science fiction, utopia, horror, and other fictions of alterity or estrangement.

Spaces and Fictions of the Weird and the Fantastic: Ecologies, Geographies, Oddities, as a whole and in each of the essays it includes, registers the profound effects of such genre fiction and media today. Significantly, editors Julius Greve and Florian Zappe have in this volume put together a sort of atlas or collection of maps by means of which readers may orient themselves with respect to the already weird and increasingly weirder geographies of our time, a moment characterized by ecological, social, political, and representational crises so severe as to defy many of the traditional means of understanding them and of conceptualizing alternatives. As Fredric Jameson has famously put it (so famously, in fact, that there's now a marvelously convoluted narrative of attribution attached to the statement), "it is easier to imagine the end of the world than to imagine the end of capitalism."[3] The dominance within popular culture of narratives explicitly invoking radically alternative realities, from science fictional scenarios and superheroes to medievalist fantasies, myths, and fairy stories, might be taken as one sign of the contemporary yearning for radically different social formations that seem beyond the reckoning of more "realistic" modes of artistic expression. Indeed, even the apocalyptic or post-apocalyptic dystopias that dominate mass culture today might be considered so many examples of a sort of utopian political unconscious at work, whereby a map of our real world that could engender alternative visions emerges from the tattered images of a destroyed environment.[4] Similarly, as the essays in the present volume demonstrate convincingly, the reemergence of weird fiction (broadly conceived) in recent years has opened up new vistas from which to view our own world and to imagine new ones.

However, as ought to be clear from the foregoing, these fantastic maps cannot be taken for accurate representations of the weird geographies they attempt to lay out before us. The maps themselves are also weird, and the spaces depicted in them are likewise fantastic. If maps have always made possible a sort of clarifying overview, it is also the case that maps have served to confuse as much as to make known, and not just in their capacity as weapons of ideological warfare. Even relatively innocent maps can produce feelings of disorientation. As Alberto Toscano and Jeff Kinkle have observed, "one of the first products of a genuine striving for orientation is disorientation, as proximal coordinates come to be troubled by wide, and at times overwhelming vistas."[5] Along those lines, weird fiction and fantasy can also prove disorienting at first, but they often leave us with empowered imaginations, which in turn allow us to see this all-too-real world in altogether different ways, making possible new maps and making visible new spaces. The contributors to *Spaces and Fictions of the Weird and the Fantastic* provide exemplary visions of how these maps operate in culture today, and this volume as a whole allows us to identify those weird and fantastic features of the contemporary world system that, like the surreal or monstrous imagery of recent weird fiction, could provide glimpses of other systems.

San Marcos, USA   Robert T. Tally Jr.

NOTES

1. Gilles Deleuze and Félix Guattari, *A Thousand Plateaus*, trans. Brian Massumi (Minneapolis: University of Minnesota Press, 1987), 12.
2. See, e.g., my *Topophrenia: Place, Narrative, and the Spatial Imagination* (Bloomington: Indiana University Press, 2019); see also my *Spatiality* (London: Routledge, 2013).
3. Fredric Jameson, "Future City," in *Ideologies of Theory* (London: Verso, 2008), 573. Much earlier, Jameson had observed that "It seems to be easier for us today to imagine the thoroughgoing deterioration of the earth and of nature than the breakdown of late capitalism; perhaps that is due to some weakness in our imagination"; see Jameson, *The Seeds of Time* (New York: Columbia University Press, 1994), xii. The notion has also been attributed to Slavoj Žižek, who in *Mapping Ideology*, had cited Jameson's comment in making a similar point. As Sean Grattan has noted, "that the phrase circulates as a somehow unattributable truism says a lot about what kinds of futures might remain unthinkable after the much heralded end of history"; see Grattan, *Hope Isn't Stupid: Utopian Affects in Contemporary American Literature* (Iowa City: University of Iowa Press, 2017), 5.
4. See, e.g., my "The End-of-the-World as World System," in *Other Globes: Past and Peripheral Imaginations of Globalization*, ed. Simon Ferdinand, Irene Villaescusa-Illán, and Esther Peeren (Palgrave Macmillan, 2019), 267–283.
5. Alberto Toscano and Jeff Kinkle, *Cartographies of the Absolute* (Winchester: Zero Books, 2015), 25.

BIBLIOGRAPHY

Deleuze, Gilles, and Félix Guattari. *A Thousand Plateaus*, trans. Brian Massumi. Minneapolis: University of Minnesota Press, 1987.

Grattan, Sean. *Hope Isn't Stupid: Utopian Affects in Contemporary American Literature*. Iowa City: University of Iowa Press, 2017.

Jameson, Fredric. "Future City." In *Ideologies of Theory*, 563–576. London: Verso, 2008.

———. *The Seeds of Time*. New York: Columbia University Press, 1994.

Tally Jr., Robert T. "The End-of-the-World as World System." In *Other Globes: Past and Peripheral Imaginations of Globalization*, edited by Simon Ferdinand, Irene Villaescusa-Illán, and Esther Peeren, 267–283. Palgrave Macmillan, 2019.

———. *Spatiality*. London: Routledge, 2013.

———. *Topophrenia: Place, Narrative, and the Spatial Imagination*. Bloomington: Indiana University Press, 2019.

Toscano, Alberto, and Jeff Kinkle. *Cartographies of the Absolute*. Winchester: Zero Books, 2015.

ACKNOWLEDGEMENTS

The editors would like to thank the following individuals without whom the exploration of *Spaces and Fictions of the Weird and the Fantastic* would not have been possible: Robert T. Tally Jr., Allie Troyanos, Rachel Jacobe, Eugene Thacker, Keith Tilford, Marleen Knipping, Susann Köhler, James Dowthwaite, Andrew S. Gross, Anca-Raluca Radu, Theresa Croll, Frederik Prush, Caro Franke, Hanna Riggert, and Jonah H. Greve. Thank you for your input, inspiration, patience, and support!

The text of Eugene Thacker's chapter, "*Naturhorror* and the Weird," has previously been published in his monograph *Tentacles Longer Than Night: Horror of Philosophy Vol. 3* (Winchester, UK: Zero Books, 2015), 143–156. The editors would like to thank John Hunt Publishing/Zero Books for generously granting permission to reprint these pages in the present book.

CONTENTS

CHAPTER 1

Introduction: Ecologies and Geographies of the Weird and the Fantastic

Julius Greve and Florian Zappe

I spread the whole earth out as a map before me.
On no one spot of its surface could I put my finger and say, here is safety.
—Mary Shelley, *The Last Man*[1]

The map had been the first form of misdirection, for what is a map but a way of emphasizing some things and making other things invisible?
—Jeff VanderMeer, *Annihilation*[2]

If there is one common conclusion to be drawn from these epigraphs—both taken from eminent examples of apocalyptic speculative fiction (albeit about two centuries apart)—it may be that *maps won't save us*. It is somewhat striking that the literary genres these two examples stem from—the Gothic and the (New) Weird—are almost obsessively preoccupied with moments

J. Greve (✉)
University of Oldenburg, Oldenburg, Germany

F. Zappe
University of Göttingen, Göttingen, Germany

© The Author(s) 2019
J. Greve and F. Zappe (eds.), *Spaces and Fictions of the Weird and the Fantastic*, Geocriticism and Spatial Literary Studies,
https://doi.org/10.1007/978-3-030-28116-8_1

1

of failure regarding the basic promises of drawing and using maps: to render the abstract reality of geography comprehensible (in spite of the well-known and much-debated representational complications ingrained in the dyad of the map and the territory) and to provide safety by reliable guidance and orientation. "On no one spot of its surface could I put my finger and say, here is safety."[3] The crisis of the map is always a marker for the collapse of an established epistemological matrix, a paradigmatic turn in the light of a new, fundamentally disruptive ontological insight, "for what is a map but a way of emphasizing some things and making other things invisible?"[4] And if we accept the claim of the late Mark Fisher that one of the defining features of the weird is its function as a "signal that the concepts and frameworks which have previously been employed are now obsolete,"[5] the recent resurgence of weird horror in popular culture may point us to the core of this shift.

Based on this assertion, this volume aims to investigate how weird fiction in literature and other narrative media can be understood as a seismographic indicator for a changing perspective on the material world surrounding us. We do not consider it a mere coincidence that a new vogue of philosophically ambitious and sophisticated speculative fiction coincided with the increasing popularization of the discourse of what has been called the Anthropocene within and beyond academia during the first decade of the twenty-first century—a time in which we had to realize that our time-honored perspectives on the world and the dynamics that shape it can no longer provide us with (the illusion of) safety and direction.

Against the backdrop of this historical moment, the present volume examines genre fiction and film—conventionally categorized as "the weird" and "the fantastic"—within the discursive framework of the environmental humanities and contemporary continental philosophy: that is, in terms of *space*, *place*, and *ecology*. The chapters in this book reflect on the convergent themes of spatiality, climate change, and related anxieties concerning the future of human affairs. They take these themes to be crucial for any understanding of current forms of weird literature and culture. Moreover, the contributions reframe central texts and contexts of this tradition, from the nineteenth to the twenty-first century, as precisely ecologically and geographically informed ways of meaning-making.

Since the US-American writer and literary critic H. P. Lovecraft introduced the concept in his essay "Supernatural Horror in Literature" (1927), the notion of "the weird" has been a haunting presence in the larger discourse on (post-)Gothic fiction—as a profitable genre label, as a category of scholarly criticism and as a distinct, yet indeterminate, mode of artistic production. Across various medial expressions of culture—literature, film, music, and beyond—both the weird and the related (and oftentimes synonymously used) conception of the fantastic has had multiple representatives: from Lovecraft and Arthur Machen to Harry Partch and the Tom Waits of *Swordfishtrombones* (1983), from Samuel R. Delany and Octavia Butler to Diane Arbus's photographs and David Lynch's transmedial oeuvre, and from Joyce Carol Oates and Jamaica Kincaid all the way to Thomas Ligotti, China Miéville, and Michael Cisco, there is a heterogeneous, seemingly unending, and ever-growing list of artists, authors, and musicians whose work may be theorized in terms of the weird and the fantastic. Especially more recent contributions such as Benh Zeitlin's movie *Beasts of the Southern Wild* (2012), the TV series *True Detective* (since 2014) or Jeff VanderMeer's Southern Reach trilogy of novels (with *Annihilation* recently adapted for the screen by Alex Garland), are just a few examples that show the increasing popularization of the genre of the weird in American culture, in particular, and the centrality of ecological and spatial concerns of the genre today. Scrutinizing the intricacies of the weird and the fantastic in relation to similar categories such as the uncanny, the grotesque, and the abject, the chapters in this volume confront—arguing from various theoretical and disciplinary perspectives—the environmental, geographical, political, and ethical contours of weird and fantastic aesthetics and poetics.

We understand weird and fantastic fiction (across all media) as a diagnostic mode of storytelling, outlining the latent anxieties and social dynamics that define a culture's "structure of feeling" (in Raymond Williams' phrase) at a given historical moment. Similar to science fiction, the cultural expression of weirdness, in many ways, seeks to sublimate or integrate that which may not be easily appropriated in the culture at large, yet which unmistakably determines the imminent future of the latter. If the genre mirrored the fear of modernization and industrialization (and its resultant social and cultural contradictions in Western society) in the first half of the twentieth century, its manifestations since the 1960s, and especially since the turn of the millennium, can be read as reflection of the creeping awareness of fundamental *ecological* and *geological* crises.[6]

In the light of this turn, we position this volume in the current and multifaceted enterprise of eco- and geocritical explorations of literature and culture by retracing an eco- and geoconscious trajectory of "the weird" in Anglophone literature that goes back to the early nineteenth century—thus, approximating the time in which some scholars have located the beginning of the Anthropocene.[7] In literary terms, the genre of the weird can be divided into (at the very least) two phases: the Old Weird—ineluctably connected to Lovecraft's poetics and politics, and to the numerous authors he lists in his essay on "Supernatural Horror"—and the New Weird—linked to a partial revision of this tradition by a highly diverse host of writers, including VanderMeer, Miéville, N. K. Jemisin and others. Interestingly enough, the formative critical perspective that seem to unite not only the Old and the New Weird on the level of form and content, but also the various topics and positions gathered in this volume may be found in Bertrand Westphal's model of a geocritical practice, as expounded in his *Geocriticism: Real and Fictional Spaces* (2007). In this book, he offers a tripartite approach to the analysis of literature and culture *tout court*, by emphasizing the notions of *spatiotemporality*, *transgressivity*, and *referentiality*. Grounded in the continental thought of precursors such as Michel Serres, Gilles Deleuze, or Henri Lefebvre, Westphal contextualizes his theory firmly in the historical period of postmodernity, recapitulating the Jamesonian argument that what time was to modernism, space becomes in postmodern culture.[8] For Westphal, however, this late twentieth-century geocritical perspective ought to give rise to an understanding of space and time in what Deleuze once called a "reciprocal determination,"[9] which actually accounts for the first principle of spatiotemporality:

> A deconstructed temporality corresponds to an explosion of spatiality, often resulting in a massive investment of geography. The semantics of tempuscules leads to an archipelagic perception of time and space. Its ideal metaphor is certainly entropy. Its universe is isotropic: the dynamic behind it does not favour any direction or configuration. Its progression defies hierarchy.[10]

Which brings us to the two additional core concepts of Westphal's approach: transgressivity and referentiality. The former is already premeditated in the violence of entropy's dissipation and, hence, violation of strict boundaries, both within the realm of biochemistry and human communication: "the dynamic behind" entropy's petrifying force "does not favour any direction" whatsoever. The latter term—referentiality—is crucial for Westphal because, even though he deems himself a theorist still operating within the cultural dominant of the postmodern, he nonetheless seeks to elide the emphasis on *textuality* in the interpretation of literature and culture. He instead opts for space and place in order to somehow return to a renewed angle on what structuralism once called "the referent"—that which has to be bracketed and perpetually disregarded. In the early twenty-first century,

> [t]he distinction between real space and represented or transposed space has blurred ... The analogy between the world as we experience it and its representation is less and less meaningful. But in any event, representation reproduces the real or, better, an experience of the real. For we must not forget that human space only exists in the modes of this experience, which, now becoming discursive, is the creator of this (geopoetic) world. Any work no matter how far from sensed reality, as paradoxical as it seems, is part of the real—and, perhaps, participates in forming the real. *Unreal realism* is therefore an option, only just *a bit more disconcerting than other postmodern processes.*[11]

Certainly, some of these thematic concerns of an "unreal realism"—reminiscent as it is of Michel Foucault's notion of an "incorporeal materialism,"[12] despite its being "a bit more disconcerting"—have already been addressed by the various branches of the environmental humanities as well as literary scholars in relation to weird and fantastic genre fiction. Donna J. Haraway's *Staying with the Trouble: Making Kin in the Chthulucene* (2016),[13] Benjamin J. Robertson's recent *None of This Is Normal: The Fiction of Jeff VanderMeer* (2018),[14] Eugene Thacker's "Horror of Philosophy" trilogy (2011–2015), and especially the first volume *In the Dust of This Planet* (2011), Brad Tabas's "Dark Places: Ecology, Place, and the Metaphysics of Horror Fiction,"[15] McKenzie Wark's *Molecular Red: Theory for the Anthropocene* (2015),[16] or Ben Woodard's *On an Ungrounded Earth: Towards a New Geophilosophy* (2013) and *Slime Dynamics* (2012),[17] undoubtedly need to be mentioned in this context, but they do not offer a synoptic view on the ecologies and geographies of weird and fantastic fiction from the nineteenth to the early twenty-first century.

Spaces and Fictions of the Weird and the Fantastic: Ecologies, Geographies, Oddities indeed *is* an attempt at mapmaking, at providing a detailed and rich cartography of the newly emerged and still uncharted territory between the discourses of genre fiction, political ecology and geography, and contemporary trends in continental theory, such as new materialism, posthumanism, and speculative realism. Scrutinizing the intricacies of the weird and the fantastic on a local and global scale, the volume negotiates the contours of these notions in terms of generic continuity and discontinuity, of spatiality and emplacement, and of ecological composition and decomposition.

Eugene Thacker's contribution on "*Naturhorror* and the Weird" maps the various territories of weird aesthetics and horror fiction from a decidedly philosophical point of view. Taking its cue from Friedrich W. J. Schelling's conception of the philosophy of nature in the nineteenth century (qua *Naturphilosophie*), he develops the notion of *Naturhorror* as "a description of 'nature' as invasive, contagious, over-running the human being like some kind of overgrown and dilapidated ruin."[18] In a second step, Thacker traces literary, scientific, and philosophical aspects of this kind of *Naturhorror* in the words of Algernon Blackwood, Kyōka Izumi, Cailín R. Kiernan, Charles Fort, China Miéville, Vilém Flusser, and H. P. Lovecraft.

Lovecraft is also the initial point of departure for Michaela Keck's essay "Uncanny New Worlds in Harriet Prescott Spofford's 'D'Outre Mort' and 'The Black Bess.'" Based on the insight that, since Lovecraft, one of the central qualities of the weird consists in its abilities to subvert and destabilize established and generally accepted ontological orders, she analyzes the unsettling experiences of space and place of the protagonists in two short stories by Harriet Prescott Spofford (1835–1921). Keck's readings of "D'Outre Mort" (1866) and "The Black Bess" (1868) expands the focus on the ominous post-human and material agencies in recent Spofford scholarship by examining in the characters' ill-fated attempts at joint homemaking in what seem to be increasingly domesticated ("D'Outre Mort") and technologically manageable ("The Black Bess") dwelling places on the one hand, but ultimately inaccessible and, therefore, unknowable territories on the other.

While Keck defines the weird through its potentials to disrupt ontologies (a quality it shares with its sibling, the Gothic), Julius Greve's chapter, "The Weird and the Wild: Media Ecologies of the Outré-Normative," goes in a different direction by asking how to think "weirdness" not only in terms of Gothic literature, but rather vis-à-vis the notion of "the wild." "Wildness"—a central concept in American literature at least from transcendentalism onward—is at the same time fundamentally different from and similar to Lovecraft's notion of "cosmic fear" summoned by the genre of weird fiction. Homing in on the ethics and aesthetics of both traditions of the weird and the wild in American culture, Greve analyzes a highly diverse host of authors and artists including Lovecraft, Henry David Thoreau, Jack London, Gary Snyder, Tzvetan Todorov, Thomas Ligotti, Werner

Herzog, John Carpenter, Wu Tsang, and Jack Halberstam. Along the lines of both Lovecraft's, Snyder's, and Halberstam's theoretical vocabularies, Greve proposes a theory of the "outré-normative" as the uneasy, yet highly productive conceptual ground on which to synthesize the discourses of weirdness and wildness.

Patricia MacCormack's chapter "Queering the Weird: Unnatural Participations and the Mucosal in H. P. Lovecraft and Occulture" explicitly returns to Lovecraft, the primordial cell of every discussion of the weird, in order to explore how the possibilities of queer-feminist readings of his works "can offer new lines of flight toward creating terrains of the weird which open spaces for anti-anthropocentric and ahuman becomings."[19] Her essay confronts the reactionary politics of the Lovecraftian manifestation of the weird not only by analyzing the concept through the critical lens of Gilles Deleuze and Félix Guattari's notion of "unnatural participations" and Luce Irigaray's work on "mucous," but also by cross-reading it with a wide spectrum of artistic representations of the weird—ranging from Katsushika Hokusai, Fernand Khnopff to Andrej Zulawski. She argues for a "queering" of our perspective by focusing of those manifestations of weird art that embrace and celebrate a subversive hybridity that "open[s] the world to strange combinations and infinite relations without hierarchy, taxonomy, classification toward an ecological plateau of queer cosmic desire."[20]

Moritz Ingwersen's "Geological Insurrections: Politics of Planetary Weirding from China Miéville to N. K. Jemisin" aims at a cartography of that recent trend in speculative fiction that conventionally goes by the name New Weird literature based on the premiss that its practitioners represent a revitalization of the weird mode that "literalizes the geomorphic enmeshments of the Anthropocene."[21] In this chapter, Ingwersen goes beyond a mere geocritical interpretation of Miéville's and Jemisin's work and rather aims to advance dialogues between theory and fiction that draw a variety of theoretical positions ranging from Eugene Thacker and Bruno Latour to Donna Haraway as they do from recent critical intervention by scholars such as Marisol de la Cadena and Vanessa Watts who confront established materialist rhetorics with Indigenous concepts such as "Place-Thought."

The (dis)continuities between the Old and the New Weird are also at the center of James Kneale's "'Indifference would be such a relief': Race and Weird Geography in Victor LaValle and Matt Ruff's Dialogues with H. P. Lovecraft." Focussing on LaValle's *Ballad of Black Tom* and Ruff's

Lovecraft Country (both from 2016), Kneale shows how the highly inter-textual works of these two writers of the New Weird revisit, rework, and remap the woeful tradition of racism in the Old Weird geographies. He argues that "these recent novels make it possible to think of white racism as itself weird or eerie, in Lovecraft's day or our present."[22]

Jolene Mathieson's chapter "The Oceanic Weird, Wet Ontologies and Hydro-Criticism in China Miéville's *The Scar*" introduces an additional and innovative geocritical concept to the discussion of New Weird fiction. Drawing on recent scholarship in the field of oceanic studies—specifically on Kimberly Peters and Philip Steinberg's concept of a decidedly non-anthropocentric "wet ontology"—Mathieson reads Miéville's celebrated 2002 novel *The Scar* as a particular mode of weird writing she calls the "oceanic weird" that "conceptualizes water, the ocean and the slime mate-rials born from it as mundane matter—but as matter whose wet ontologies are so radical, so alien, that they enact and constitute a *hypermateriality*—a materiality that touches and configures the human but, paradoxically, can seemingly only be accessed speculatively through the metonymic and metaphoric paradigm of the mythological and the unreal."[23] Her inter-pretation of Miéville identifies the maritime biosphere a sthe sphere o f an immanent and material, yet somewhat impalpable uncanniness (which has, as her essay also demonstrates, a long tradition in Western culture) that cer-tainly differs from yet also inevitable recalls the unknowable cosmic worlds of the Old Weird tradition as a space of nonhuman alterity.

Gry Ulstein explores another seminal text of the New Weird canon, Jeff VanderMeer's Southern Reach trilogy, with a focus on its potentials to shift established coordinates of spatial thinking in the current age. Based on Rebecca Solnit's claim that "[h]ope is an embrace of the unknown and the unknowable," this chapter, titled "'Through the eyes of Area X': (Dis)locating Ecological Hope via New Weird Spatiality," explores an emer-gent hopeful tone in New Weird narrative as one of the most important contrasts to the traditional (or Old) weird: The New Weird's intrusion of the global into the local, of monstrosity into normality productively dis-turbs the human concept of space and challenges readers to resituate them-selves sustainably in the increasingly weirded time of the Anthropocene.

Ben Woodard's "Inexistent Ink: Michael Cisco and Quentin Meillassoux on Writing Worlds" reads Cisco's works against the backdrop of Meillas-soux's semiotics, especially of the philosopher's notion of the "sign devoid of meaning." Focusing on Cisco's novels *The Divinity Student* (1999),

Member (2013) and *Unlanguage* (2018) as case studies, Woodard investigates how "the weird touches on the oblique construction that accompanies the narrative matter of text itself," an analysis that does not limit itself to the inscription as "material affect" but rather as "as imperfect index of another world."[24]

Marius Henderson's "Notes on the Alluring Weirdness of (Materialist) Rumination and Regurgitation: Reading Ariana Reines and Jamie Stewart" takes Ariana Reines's poetry collection *The Cow* (2006) as well as specific aspects of the sonic, lyrical, and visual works of Jamie Stewart/Xiu Xiu as touchstones (both of which make ample use of montage-like juxtapositions) and aims to test possible relational connections between weird (politico-) aesthetic practices of rumination and regurgitation and contemporary materialisms, such as Elizabeth Wilson's new materialist "gut feminism" (2015). These artistic and theoretical approaches are likewise devoted to the juxtaposition of the visceral, abdominal, and gustatory with its ostensible antonyms and encounter dynamics of abstraction inherent to capitalism, like the emergence of abstract labor and the commodity-form, by proliferating modes of weirdness in terms of both visual and auditory prosody.

The final chapter in this volume, Marlon Lieber's "Spaces of Communal Misery: The Weird Post-Capitalism of *Beasts of the Southern Wild*," engages with Benh Zeitlin's 2012 film *Beasts of the Southern Wild*, which has been criticized for its stereotypical representation of black poverty. Following Fredric Jameson's claim that mass cultural products articulate "reification" and "utopian" impulses, Lieber critically analyzes the film's ambivalent politics. On one hand, he argues that Zeitlin's formal investment in the beauty of the nonnormative culture of the "beasts" leads to an affirmation of abject poverty in the guise of demanding recognition for the abject. Thus, the film—whose plot centers around a storm reminiscent of Hurricane Katrina—suggests that any form of state action meant to protect vulnerable populations equals an illegitimate biopolitical intervention. By taking that position, the film embraces a radical libertarian ideology that rejects any form of government intervention even if this means having to accept extreme human misery—which here is disavowed by turning the humans into "beasts" well-equipped for survival in the spaces of the "Southern Wild." On the other hand, the film is committed to expressing utopian desires by representing a community that is not structured by capitalist social relations—the "beasts" do not engage in wage labor in order to live, and their interactions are not mediated by the money-form. Ultimately, Lieber argues, the film may be read as a symptom of the seeming impossibility to imagine noncapitalist social relations except in the form of an isolated and immiserated community/commune subject to nature's whims and unable to expand.

Given the equally diverse and homogeneous nature of these contributions, it needs to be said, finally, that, in accordance with the two epigraphs with which we started, it is the objective of this book to demonstrate the multiple ways in which weird and fantastic literature may be figured and theorized, and how the essential disposition of a geologically and ecologically informed concept of the weird is key in any understanding of the latter. Mapping the weird, it seems, is especially pertinent, in times that put to question the relevance and adequacy of maps. Unreal realism, speculative fiction, place-thought, and wet ontology: ecologies and geographies of the weird and the fantastic continue to be, and always have been, as it were, blessed by "the angel of the odd."[25]

NOTES

1. Mary Shelley, *The Last Man* (Mineola: Dover Publications, 2010), 188.
2. Jeff VanderMeer, *Annihilation* (New York: Farrar, Straus and Giroux, 2014), 66.
3. Shelley, *The Last Man*, 188.
4. VanderMeer, *Annihilation*, 66.
5. Mark Fisher, *The Weird and the Eerie* (London: Repeater Books, 2016), 13.
6. Eugene Thacker, *In The Dust of This Planet: Horror of Philosophy*, Vol. 1 (Winchester: Zero Books, 2011).
7. See, for instance, Jason Moore, *Capitalism in the Web of Life: Ecology and the Accumulation of Capital* (London: Verso, 2015). For an important recontextualization of "anthropocenic origins" and their exploitative politics, see chapter 2 of Kathryn Yusoff's *A Billion Black Anthropocenes or None* (Minneapolis: University of Minnesota Press, 2018).
8. Bertrand Westphal, *Geocriticism: Real and Fictional Spaces*, trans. Robert T. Tally Jr. (Basingstoke: Palgrave Macmillan, 2011), 2.
9. Gilles Deleuze, *Difference and Repetition*, trans. Paul Patton (London: Continuum, 2004), 57, 220–221.
10. Westphal, *Geocriticism: Real and Fictional Spaces*, 20.
11. Ibid., 84–85; emphasis added.
12. Michel Foucault, *The Archaeology of Knowledge & The Discourse on Language*, trans. A. M. Sheridan Smith (New York: Pantheon Books, 1972), 230.
13. Donna J. Haraway, *Staying with the Trouble: Making Kin in the Chthulucene* (London: Duke University Press, 2016).

14. Benjamin J. Robertson, *None of This Is Normal: The Fiction of Jeff Vander-Meer* (Minneapolis: University of Minnesota Press, 2018).
15. Brad Tabas, "Dark Places: Ecology, Place, and the Metaphysics of Horror Fiction," *Miranda* 11 (July 2015): 1–17.
16. McKenzie Wark, *Molecular Red: Theory of the Anthropocene* (London and New York: Verso, 2015).
17. Ben Woodard, *On an Ungrounded Earth: Towards a New Geophilosophy* (Brooklyn, NY: Punctum, 2013) and *Slime Dynamics* (Winchester, UK: Zero Books, 2012).
18. See Thacker in this volume, 15.
19. See MacCormack in this volume, 57.
20. Ibid., 69.
21. Ingwersen in this volume, 74.
22. Kneale in this volume, 95.
23. Mathieson in this volume, 114.
24. Woodard in this volume, 149.
25. We take this phrase from Edgar Allan Poe's short story by the same name, in *Spirits of the Dead: Tales and Poems* (London: Penguin, 1997), 223–232.

Bibliography

Deleuze, Gilles. *Difference and Repetition.* Translated by Paul Patton. London: Continuum, 2004.

Fisher, Mark. *The Weird and the Eerie.* London: Repeater Books, 2016.

Foucault, Michel. *The Archaeology of Knowledge & The Discourse on Language.* Translated by A. M. Sheridan Smith. New York: Pantheon Books, 1972.

Haraway, Donna J. *Staying with the Trouble: Making Kin in the Chthulucene.* London: Duke University Press, 2016.

Moore, Jason. *Capitalism in the Web of Life: Ecology and the Accumulation of Capital.* London: Verso, 2015.

Poe, Edgar Allan. *Spirits of the Dead: Tales and Poems.* London: Penguin, 1997.

Robertson, Benjamin J. *None of This Is Normal: The Fiction of Jeff VanderMeer.* Minneapolis: University of Minnesota Press, 2018.

Shelley, Mary. *The Last Man.* Mineola: Dover Publications, 2010.

Tabas, Brad. "Dark Places: Ecology, Place, and the Metaphysics of Horror Fiction." *Miranda* 11 (July 2015): 1–17.

Thacker, Eugene. *In The Dust of This Planet: Horror of Philosophy,* Vol. 1. Winchester: Zero Books, 2011.

VanderMeer, Jeff. *Annihilation.* New York: Farrar, Straus and Giroux, 2014.

Wark, McKenzie. *Molecular Red: A Theory of the Anthropocene.* London and New York: Verso, 2015.

Westphal, Bertrand. *Geocriticism: Real and Fictional Spaces*. Translated by Robert T. Tally Jr. Basingstoke: Palgrave Macmillan, 2011.

Woodard, Ben. *On an Ungrounded Earth: Towards a New Geophilosophy*. Brooklyn, NY: Punctum, 2013.

———. *Slime Dynamics* Winchester: Zero Books, 2012.

Yusoff, Kathryn. *A Billion Black Anthropocenes or None*. Minneapolis: University of Minnesota Press, 2018.

CHAPTER 2

Naturhorror and the Weird

Eugene Thacker

NATURHORROR

Near the end of the eighteenth century, the German philosopher, novelist, and aphorist Friedrich Schelling sought to formulate a renewed philosophy of nature that would combine the findings of the physical sciences with that of speculative philosophy in the wake of Immanuel Kant. Kant's critical philosophy had forged a wedge between the world in itself and our perception of the world, but post-Kantian thinkers like Schelling sought ways of bridging Kant's division of self and world. For Schelling, the key intuition was that the self that thinks about the world is also part of the world, and it is a mistake to presume that there is first a separately existing self that then turns toward and reflects on the world as an object. The world that the self thinks about is also "in" the self, and the two share something in common that may not be reducible to either.

What, then, is that "something" that is common to both self and world? For Schelling, "nature" came to signify that commonality—but by nature Schelling does not mean "the outdoors" or forests and oceans, and neither does he mean nature in the sense of a fixed essence (as in "human nature").

E. Thacker (✉)
The New School, New York, NY, USA

© The Author(s) 2019
J. Greve and F. Zappe (eds.), *Spaces and Fictions of the Weird and the Fantastic*, Geocriticism and Spatial Literary Studies,
https://doi.org/10.1007/978-3-030-28116-8_2

Nature for Schelling that "something" that was not in itself anything, a unity without boundary constituted entirely of process, becoming, flux and flow—a "ground" of the world that was also continually ungrounding itself. In his study *Philosophies of Nature After Schelling*, Iain Hamilton Grant encapsulates this intuition: "Self-conscious subjectivity, therefore, is simply the 'highest power' of the 'identity of subjective and objective we call nature.'"[1] The stakes of such a philosophy are high, as Grant summarizes:

> If, in other words, "to philosophize about nature means to create nature," the latter cannot be a nature restricted apriori by the particular physiological means by which it philosophizes. Instead, nature philosophizing must itself be unconditioned, so that the range of instances of natural philosophy must extend beyond the remit of physiologically conditioned particulars such as species, or even phylla.[2]

In his experimental and syncretic approach, Schelling incorporated ideas from physics, biology, geology, chemistry, as well as elements from mystical theology and classical myth. Given the questions Schelling is posing, such a philosophy would no longer simply be a philosophy *of* nature, as if the philosopher stood above and apart from nature. For if the nature that I am thinking about is also in me and coursing through me, then it follows that I am in some way identical with nature, and that nature "thinks" me just as I think nature. As Grant puts it, "what thinks in me is what is outside me." Instead of a philosophy of nature, Schelling proposes a *naturphilosophie*. Thought becomes strangely impersonal and nonhuman, just as the human subject becomes at once the one who thinks nature and that which is thought by and through nature.

But if this is the case, and what thinks in me is also outside me, it is also possible that what is outside me is also outside of my individual, subjective concerns, hopes, and desires, just as what thinks in me is also alien, impersonal, and similarly estranged to the "me" that is thought. My thoughts are not my own, and I am thought by an enigmatic nonentity whose motives may be contrary to my own—or which may have no motives at all. Contra Schelling's romantic conception of the unity of self and world, the nature in me and the nature outside me, there is another kind of "nature," one that courses through self and world, but that does so without aim or end, indifferent to the self's possessive individualism and the species' sense of superiority. What thinks in me is what is outside me. Is this not also a description of "nature" as invasive, contagious, over-running the human being like some kind of overgrown and dilapidated ruin? Would this not transform Schelling's *naturphilosophie* into a *naturhorror*?

Among authors of supernatural horror, no one has examined the role of nature more than Algernon Blackwood. In story after story Blackwood's characters discover a nature that is a darker version of Schelling's *naturphilosophie*, and nature that they are, at the same time, drawn to, often undergoing ambivalent transformations. Lyrical and foreboding, Blackwood's story "The Man Whom the Trees Loved" (1912) tells of a husband and wife enjoying their cabin in the woods. Gradually the husband becomes more and more interested in the trees that surround their house, taking long walks in the forest, himself becoming more and more silent and preoccupied. He dreams of the slow, swaying, forest trees, and his dreams begin to blur into his daily life. The couple can't tell, but it seems as if the trees outside their house have moved ever so imperceptibly closer. At night, amid the "strong smell of mould and fallen leaves" the wife feels a presentment of something close but intangible, something linked to the trees outside: "The horrible, dim enchantment of the trees was close about them in the room—gnarled, ancient, lonely trees of winter, whispering round the human life they loved."[3] One night, amid stirring dreams, the wife suddenly awakes. There is dew and wet leaves on her clothing, as if she had been outside. In the dim, moonlit darkness, she glances over at her husband, sound asleep. Then she looks up:

> …what caught her unawares was the horrid thing that by this fact of a sudden, unexpected waking she had surprised these other things in the room, beside the very bed, gathered close about him while he slept…She screamed before she realised what she did—a long, high shriek of terror that filled the room, yet made so little actual sound. For wet and shimmering presences stood grouped all round that bed. She saw their outline underneath the ceiling, the green, spread bulk of them, their vague extension over walls and furniture. They shifted to and fro, massed yet translucent, mild yet thick, moving and turning within themselves to a hushed noise of multitudinous soft rustling. In their sound was something very sweet and winning that fell into her with a spell of horrible enchantment.[4]

Seamlessly crossing nature outside and nature inside, waking and dreaming, the animate and inanimate, the eerie forest trees in Blackwood's story literally envelop the characters in an intimate, lulling terror, a "horrible enchantment" that pulls the human characters further and further away from the human.

Something similar happens in Kyōka Izumi's story "The Holy Man of Mount Kōya" (1900), which tells the story of a young Buddhist monk making his way to a monastery near the mountain, and the various events that occur along the way. The monk begins his journey on a road, which then turns to a path, which then vanishes, leaving the monk directionless as he walks into a dense forest. He sees a snake, various insects, and at one point discovers a leech on his arm. Then, inexplicably, a torrent of leeches descends upon him from the trees. Fatigued, delirious, and panic-stricken, "it was then," the monk states, "that the strangest thought occurred to me": "These terrifying mountain leeches had been gathered there since the age of the gods, lying in wait for passersby...And at the same time, all these enormous trees, large enough to block out even the midday sun, will break into small pieces that will then turn into even more leeches..."[5] Foreboding in its appearance, the forest through which the monk walks literally takes on animistic properties, the boundary between vegetable, animal, and mineral passing away. Recalling the experience later, the monk solemnly observers:

> The destruction of mankind will not come with the rupture of the earth's fragile crust and with fire pouring down from the heavens. Nor will it come when the waves of the ocean wash over the land. Rather, it will begin with the forests of Hida turning into leeches and end with black creatures swimming in blood and muck. Only then will a new generation of life begin.[6]

This intermingling of forms is at the center of a group of short stories by the contemporary horror author and paleontologist Caitlín R. Kiernan. Published in her 2000 collection *Tales of Pain and Wonder*, each of the stories focuses on a particular place where some undefined scientific anomaly is accidentally discovered. In the story "In the Water Works (Birmingham, Alabama 1888)," it is the construction and mining site near Red Mountain, at the tip of the Appalachias. There, geology teacher Henry S. Matthews carries out his research under the auspices of the Birmingham Water Works and the Elyton Land Companies, gathering samples, accumulating data, a northerner minding his own business down south. Eventually the miners discover what appears to be a massive fault within the mountain. The fore-man asks Matthews to come with them to take a look. As they go deeper and deeper into the mountain, Matthews notices a strange, repulsive, rotten stench, the stench, he thinks of dark, fungoid, mold-ridden rocks. When they finally arrive at the breach, Matthews, still repulsed by the smell, isn't sure what he sees: "...a wide crevice in the wall of the tunnel maybe four feet across and dropping suddenly away into darkness past the reach of the lantern, running west into more blackness but pinching closed near the tunnel's ceiling."[7] The foreman holds Matthews' arm as he slowly leans over into the blackness. He discovers the bottom of the cavern is flooded, the black water infinitely deep. Looking closer, the foreman still holding him, Matthews notices something else:

> At first he doesn't see anything, angle a little less than ninety degrees where black rock meets blacker water, and then he does see something and thinks it must be the roots of some plant growing in the pool, or, more likely, runing down from the forest above to find this hidden moisture. Gnarled roots as big around as his arm, twisted wood knotted back on itself...But one of them moves, then, abrupt twitch as it rolls away from the others...[8]

Matthews watches in dismay as one of the roots rises out of the water, and, dripping, glides slowly toward him—and on its underside a small, fossil-like worm, "coiling and uncoiling, and here are a thousand of them, restless polyps sprouting from this greater appendage, row upon writhing row, and now it's risen high enough that the thing is right in front of him, shimmering in the lantern light, a living question mark..."[9]

That question mark is at the center of both the horror genre and philosophy. It demands to be taken both literally and figuratively. And it is a question mark posed to the sciences as well. This is something that American author and purveyor of the paranormal Charles Fort knew well. In a string of books published between 1919 and 1932, Fort, spending hours upon hours in the New York Public Library, amassed material on unexplained phenomena. Scientific papers, philosophy books, news reports, travelogues, even anecdotal accounts—nothing escaped from Fort's keen eye, as he gathered data on those phenomena which had been excluded from the existing scientific and philosophical systems. Some of his books gathered together data relating to biological anomalies, while others pertained to geological anomalies, and still others to cosmological anomalies. At the center of his project lay a deep interest in the limits of human knowledge:

> All attempted organizations and systems and consistencies, some approximating far higher than others, but all only intermediate to Order and Disorder, fail eventually because of their relations with outside forces. All are attempted completenesses. If to all local phenomena there are always outside

forces, these attempts, too, are realizable only in the state of completeness, or that to which there are no outside forces.[10]

Such phenomena were that portion of knowledge, that which must be excluded in order for systematic knowledge and natural laws to exist at all. His books read less like coherent theories and more like a compendium of facts and figures put in narrative form, the "data of the damned." A deep-seeded skepticism undergirds Fort's books: "Nothing has ever been defined. Because there is nothing to define."[11] Or: "Nothing has ever been finally found out. Because there i snothing to find out."[12] Or again: "We are not realists. We are not idealists. We are intermediatists—that nothing is real, but that nothing is unreal: that all phenomena are approximations one way or the other between realness and unrealness."[13] What Fort attempted in works like *The Book of the Damned* was nothing less than a "procession of data that Science has excluded." At the same time, Fort's books are mostly absent of any conclusions, his authorial voice in the selection and juxta-position of "damned" facts, as if "the facts themselves" revealed things far more horrific and unmentionable than the fancies of a writer's imagination.

An Exegesis on Tentacles

There is never just one tentacle, but many. And yet, the many tentacles always seem to trail off into nothing, into a distant ocean abyss as black as the ink it secretes. The cephalopod occupies this duality, a multiplicity of seemingly incongruous features—tentacles and multiple "arms" with suckers, a razor-sharp "beak," a complex nervous system, rows of intestinal "teeth," and a formless "head"—whose coherence falls apart once one tries to make sense of the whole creature. Emerging from a lightless ocean depth, the tentacles seem to lead back down to the abyss from which it came, a multiplicity dissipating into a slumberous, slow, and alien depth. When the tentacles are not reaching down, they are reaching up, for prey— fish, crustaceans, a whale, a sea-borne ship, a flailing human body. This is the cultural mythos, at least. From medieval Icelandic fables of the Kraken, to Jules Verne's *20,000 Leagues Beneath the Sea*, tentacles envelop human beings in their unhuman embrace, the abyss of the unknown sea reaching up to the surface with a certain inevitability.

At the same time, cephalopods are products of human knowledge-production. We have given them their various names, both scientific and colloquial, and they make appearances in the natural history works of Pliny the Elder and Linnaeus, as well as in the Teuthological studies of modern marine biologists. This has not, however, prevented the periodic and often dubious reports of giant squids, which continue to this day. At once products of science and myth, a multiplicity receding into an abyss, an alien creature become like a god—these are themes woven into China Miéville's 2010 novel *Kraken*, his own rendition of cephalopod-lore.

Like many of Miéville's novels, the backdrop of *Kraken* is urban London. But counterposed to this is the vast sea. One is bustling with noise, the other enigmatically silent. One is built up or broken down, the other flows, sometimes turbulently and sometimes imperceptibly. In the interzone between city and sea *Kraken* introduces us to a whole bestiary of groups and organizations—the London Natural History Museum, the Darwin Centre, the FSRC (Fundamentalist and Sect-Related Crime Unit), the Krakenist cult, Londonmancers, the Brotherhood of the Blessed Flood, Knuckle-heads, Gunfarmers, Chaos Nazis, and other nefarious grifters, gangsters, and cultists, each warring for their own apocalypse. Caught in the middle of it all is Billy Harrow, ex-student and curator at the Natural History Museum. But running through everything is the enigmatic and mute figure of the giant cephalopod, encased in museum glass as visitors gaze at it with an equal muteness.

All-touching, all-seeing, an alien intelligence writing with an ink illegible but in the sea, in *Kraken* the cephalopod becomes at once an object of religion and of science—an object of religion because of science. Beneath the all-too-human veneer of urban London patterns emerge, linking city, sea, and the unhuman forms of life that pass between them. As one character notes, "What was squiddity but otherness, incomprehensibility. Why would such a deity understand those bent on its glory? Why should it offer anything? Anything at all?"[14] An old Londonmancer, deciphering the bloody entrails of a cut-open city concrete, gives a foreboding reading: "Everything closing down. Something coming up."[15]

But if, for these fringe groups, the cephalopod has become a god, it is not an anthropomorphic or even communicative god. In the library of the Krakenist cult, in a concrete bunker deep underground, Billy discovers a storehouse of "cephalopod folklore"—*Moby Dick*, Jules Verne, poems by Hugh Cook and Tennyson, obscure scientific papers, and secret treatises such as the dreaded *Apocrypha Tentacula*. In one of the books he reads:

> We cannot see the universe. We are in the darkness of a trench, a deep cut,
> dark water heavier than earth, presences lit by our own blood, little biolumes,

heroic and pathetic. Promethei too afraid or weak to steal fire but able to still glow. Gods are among us and they care nothing and are nothing like us. This is how we are brave: we worship them anyway.[16]

In *Kraken*, the alterity of the cephalopod is matched only by its indifference toward the various groups that attempt to interpret it, and thus to control it. Eventually the enigma of the cephalopod becomes the enigma of the sea itself. At one point in the novel Billy visits the Brotherhood of the Blessed Flood, who protect an oracle of the sea. But the sea remains impervious to human demands. Billy muses:

> The sea is neutral. The sea didn't get involved in intrigues, didn't take sides in London's affairs. Wasn't interested. Who the hell could understand the sea's motivations, anyway? And who would be so lunatic as to challenge it? No one could fight that. You don't go to war against a mountain, against lightning, against the sea. It had its own counsel, and petitioners might sometimes visit its embassy, but that was for their benefit, not its. The sea was not concerned: that was the starting point.[17]

The starting point is, then, this indifference of the world. It is an insight found not only in myths and fables but in the treatises of natural history and science. Commenting on the myriad of monsters in the sea, physician Ambroise Paré notes, in his sixteenth-century *On Monsters and Marvels* (*Des Monstres et Prodiges*), that "awesome things likewise happen on water. For huge flames of fire spreading across the water have been seen issuing from the abysses and whirlpools of the sea—a very monstrous thing—as if the great quantity of water could not stifle the fire; in this God shows himself incomprehensible, as in all his works."[18]

Whether one puts it in the language of fiction or of science, the result is the same—the sudden realization of a stark, "tentacular" alienation from the world in which one is enmeshed. For these and other texts the cephalopod stands in as a manifestation of that indifference of the black, inky abyss. And in a modern context it is largely through scientific knowledge-production that this tentacular alienation is made possible. We human beings are, after all, the ones who have dreamed up taxonomies, anatomies, and nomenclatures, making possible the most systematic, rigorous articulation of this alienation. In a sense, the result of scientific classification is not that we as human beings finally find our place in the world, but the reverse—that we increasingly feel ill-at-ease in the world. It is we who are alien.

This is the insight of *Vampyroteuthis Infernalis*, an incomparable and strange work that straddles the line between science and fiction, biology, and horror. Published in the 1980s by the philosopher and journalist Vilém Flusser, the *Vampyroteuthis* is written in the language of a biological classification textbook or a marine biology research paper. Appended to it is a report by the Scientific Institute for Paranaturalist Research, a set of anatomical illustrations and accompanying text, both by the artist Louis Bec. The text details the purported discovery of a new species of hyper-complex, "intelligent" cephalopod, dubbed *Vampyroteuthis infernalis* due to its many rows of undulating teeth and the general "voracity of their expression." As a whole, the *Vampyroteuthis* text is a study in self-reflexivity; the language of biological classification is not simply mocked but respectfully, even enthusiastically employed, to give the reader a sense of the strangeness of much of life on the planet. The cephalopod is the key figure in the text, at once the most remote from the human being, at yet, as Flusser contends, at the same time uncannily near us:

> ...the vampyroteuthis is not entirely alien to us. The abyss that separates us is incomparably smaller than that which separates us from extraterrestrial life...We are pieces of the same game, both constructed of genetic information, and we belong to a branch of the same phylogenetic tree to which its branch belongs. Our common ancestors dominated the beaches of the earth for millions of years, and it was relatively late in the history of life that our paths began to diverge...[19]

For all its playfulness, at the center of the *Vampyroteuthis* report is this theme of relating to a form of life radically unhuman. As Flusser notes "without any previous knowledge of biology, we feel a sense of belonging to our phylum whenever we step on a mollusk, on the one hand, or when we hear, on the other, a crackling bone under our shoe. We feel a connection with life-forms supported by bones, while other forms of life disgust us." This leads Flusser to the evocative idea of a "biological existentialism": "Though existential philosophy has concerned itself with the idea of disgust, it has never attempted to formulate a category of 'biological existentialism,' to advance something like the following hypothesis: 'Disgust recapitulates phylogenesis.'"[20]

Disgust recapitulates phylogenesis. Here Flusser is playing on the truism of genetics "ontogeny recapitulates phylogeny," which argues that the development of the individual organism replays the evolutionary development of its species. Flusser applies this to the affective domain of disgust, so beloved by existentialist thinkers such as Sartre and Camus. Later in the text Flusser gives a more tangible definition: "The more disgusting something is, the further removed it is from humans on the phylogenetic tree."[21] The insight of the science of biological classification is, Flusser suggests, to have articulated in great detail the scope and range of human disgust toward that which is not human.

Such disgust is also, of course, a sign of the human being's failure to appreciate or relate to that which is not human, other than in terms of utility for us as human beings. This is especially evident in the many horror tales featuring cephalopods, kraken, giant squids, and other nefarious sea creatures, from the *Orlando Furioso* to the tales of nautical horror by the likes of Edgar Allan Poe and William Hope Hodgson. A case in point is the H. P. Lovecraft short story "Dagon," published in 1919. The plot of the story is threadbare, and found in many of Lovecraft's tales—an unnamed narrator tells of his being lost in a strange and terrifying place, and of the unmentionable things he sees there, which, of course, no one believes once he returns home, driving him to the brink of madness. Disgust pervades the story, from the narrator's first impressions of the strange "island": "When at last I awaked, it was to discover myself half sucked into a slimy expanse of hellish black mire which extended about me in monotonous undulations as far as I could see, and in which my boat lay grounded some distance away... The region was putrid with the carcasses of decaying fish, and of other less describable things which I saw protruding from the nasty mud of the unending plain."[22] Lost without compass or map, shipwrecked and delirious, the impression leads the narrator into the depths of cosmic horror: "Perhaps I should not hope to convey in mere words the unutterable hideousness that can dwell in absolute silence and barren immensity."[23]

Though the island appears uninhabited, the remnants of what he does discover makes no sense within the ambit of existing human history, archeology, and science. It is unclear if the "island" is a recent upheaval of rock from a deep-sea volcano. A deep ravine leads to a dark pool and a giant black, "Cyclopean monolith," covered in strange symbols and depictions of unrecognizable creatures, all unmistakably sculpted by some unnamed and unseen entity.

"Then suddenly I saw it." Something "loathsome" rising to the surface of dark waters, sliding into partial view, a "stupendous monster of nightmares," "gigantic scaly arms" reaching around the monolith, a "hideous" and bowed head, emitting "certain measured sounds." "I think I went mad then."

The narrator's frantic escape and rescue does not console him. He is tainted with an anthropocentric disgust that will not leave him, and the only thing he is disgusted by more than cephalopod-like creatures from the deep is his own species, unable to comprehend a world both alien and indifferent to the human perspective:

> I cannot think of the deep sea without shuddering at the nameless things that may at this very moment be crawling and floundering on its slimy bed, worshipping their ancient stone idols and carving their own detestable likenesses on submarine obelisks or water-soaked granite. I dream of a day when they may rise above the billows to drag down in their reeking talons the remnants of puny, war-exhausted mankind—of a day when the land shall sink, and the dark ocean floor shall ascend amidst universal pandemonium.[24]

His final words at once a prayer and indictment.

Notes

1. Iain Hamilton Grant, *Philosophies of Nature After Schelling* (London: Continuum, 2006), 169. For an examination of the links between Schelling's philosophy and modern science fiction and horror, see Ben Woodard's, *On an Ungrounded Earth: Towards a New Geophilosophy* (Brooklyn: Punctum, 2013).
2. Ibid., 2.
3. Algernon Blackwood, "The Man Whom the Trees Loved," in *Ancient Sorceries and Other Weird Stories*, ed. S. T. Joshi (New York: Penguin, 2002), 268, 271.
4. Ibid., 268.
5. Izumi Kyōka, "The Holy Man of Mount Kōya," in *Japanese Gothic Tales*, trans. Charles Shirō Inouye (Honolulu: University of Hawaii Press, 1996), 36–37.
6. Ibid., 37.
7. Caitlín R. Kiernan, "In the Water Works (Birmingham, Alabama 1888)," in *Tales of Pain and Wonder* (Atlanta: Meisha Merlin Publishing, 2001), 294.
8. Ibid., 296.
9. Ibid.
10. Ibid., 9.
11. Ibid., 13.
12. Ibid., 14.
13. Ibid.

14. China Miéville, *Kraken* (London: Pan Books, 2011), 417.
15. Ibid., 186.
16. Ibid., 106. The passage is adapted from Rondelet's fifteenth-century teratology treatise *Crinis Abyssi*.
17. Ibid., 295.
18. Ambroise Paré, *On Monsters and Marvels*, trans. Janis L. Pallister (Chicago: University of Chicago Press, 1982), 161.
19. Vilém Flusser and Louis Bec, *Vampyroteuthis Infernalis: A Treatise, with a Report by the Institut Scientifique de Recherche Paranaturaliste*, trans. Valentine A. Pakis (Minneapolis: University of Minnesota Press, 2012), 6.
20. Ibid., 11.
21. Ibid.
22. H. P. Lovecraft, "Dagon," in *The Call of Cthulhu and Other Weird Stories*, ed. S. T. Joshi (New York: Penguin, 1999), 1–2.
23. Ibid., 2.
24. Ibid., 6.

BIBLIOGRAPHY

Blackwood, Algernon. "The Man Whom the Trees Loved." In *Ancient Sorceries and Other Weird Stories*, edited by S. T. Joshi. New York: Penguin, 2002.

Flusser, Vilém, and Louis Bec. *Vampyroteuthis Infernalis: A Treatise, with a Report by the Institut Scientifique de Recherche Paranaturaliste*. Translated by Valentine A. Pakis. Minneapolis: University of Minnesota Press, 2012.

Grant, Iain Hamilton. *Philosophies of Nature After Schelling*. London: Continuum, 2006.

Kiernan, Caitlín R. *Tales of Pain and Wonder*. Atlanta: Meisha Merlin Publishing, 2001.

Kyōka, Izumi. "The Holy Man of Mount Kōya." In *Japanese Gothic Tales*, translated by Charles Shirō Inouye. Honolulu: University of Hawaii Press, 1996.

Lovecraft, H. P. "Dagon." In *The Call of Cthulhu and Other Weird Stories*, edited by S. T. Joshi. New York: Penguin, 1999.

Miéville, China. *Kraken*. London: Pan Books, 2011.

Paré, Ambroise. *On Monsters and Marvels*. Translated by Janis L. Pallister. Chicago: University of Chicago Press, 1982.

Woodard, Ben. *On an Ungrounded Earth: Towards a New Geophilosophy*. Brooklyn: Punctum, 2013.

CHAPTER 3

Uncanny New Worlds in Harriet Prescott Spofford's "D'Outre Mort" and "The Black Bess"

Michaela Keck

At the heart of H. P. Lovecraft's weird fiction lies the staging of ontological confusion. Representations of immense, fantastic spaces and more-than-human powers confound the human sense of time and space, as well as the human place in the universe. At times when anthropogenic activities have become unprecedentedly powerful geological forces, the renewed interest in Lovecraft's unnerving creatures, spaces, and cosmic phenomena is hardly surprising, as his notion of the weird meets with our present critical and popular cultural occupation with environmental crisis, human and non-human ontology, and material agency. Similar to gothic literature, weird fiction involves unexplainable, unsettling encounters with the unknown. Yet, in contrast to gothic literature, weird fiction generally veers away from common notions of the supernatural. It derives its disorienting horror from, as Lovecraft put it, the "outer, unknown forces" of the universe and the

M. Keck (✉)
University of Oldenburg, Oldenburg, Germany

© The Author(s) 2019
J. Greve and F. Zappe (eds.), *Spaces and Fictions of the Weird and the Fantastic*, Geocriticism and Spatial Literary Studies, https://doi.org/10.1007/978-3-030-28116-8_3

"suspension or defeat of those fixed laws of Nature which are our only safe-guard against the assault of chaos and the daemons of unplumbed space."[1] Hence, the weird tends to frustrate transcendental explanations and spiritual speculations.

It was Edgar Allan Poe, whom Lovecraft granted the most prominent position in the nineteenth-century American genealogy of weird fiction. With the exception of Mary E. Wilkins ("The Shadows on the Wall") and Charlotte Perkins Gilman ("The Yellow Wallpaper"), Lovecraft's canon constituted an exclusively white, male-dominated tradition.[2] With this contribution, I wish to shed light on two lesser-known stories of the nineteenth-century American writer and poet Harriet Prescott Spofford (1835–1921), whose stories—like Poe's—are on the cusp of "a literature of cosmic fear."[3] Scholars recovered Spofford's many diverse writings in the wake of second-wave feminism, appraising her short fiction in the (female) canons of the American Gothic and the American Renaissance. Since the 1990s, the focus has shifted to the ecocritical, post-humanist, and new materialist aspects of her most widely read short stories. Scholars such as Ian Marshall, Dana Luciano, Matthew Sivils, or Jeffrey Andrew Weinstein have explored the ways in which the material and animal agency, as well as the geological deep time in her stories fundamentally unsettle, even threaten, human life and being in the world. I wish to expand this more recent scholarly focus by examining the weird experiences of Spofford's protagonists, as they move through space and place in "D'Outre Mort" (*The Galaxy*, November 1866) and "The Black Bess" (*The Galaxy*, May 1868). I am particularly interested in the characters' ill-fated attempts at joint homemaking in what seem to be increasingly domesticated ("D'Outre Mort") and technologically manageable ("The Black Bess") spaces, on the one hand, but still terrifying environments, on the other. Historically and discursively, Spofford places her protagonists at the intersections of gender, empire, and nature ("D'Outre Mort"), as well as gender, technology, and medical science ("The Black Bess"). Aesthetically and stylistically, she imbues the worlds of her fictitious characters with a baroque vitalism and more-than-human agency, which draws on yet also undermines Romantic notions of transcendence and highlights moments of existential crisis.

"D'OUTRE MORT": COSMIC SPACE, MYTH, AND THE HORRORS OF AUTOPOIESIS

In "D'Outre Mort," Spofford renders the tale of the forfeited love between Orient and Reymund. Visiting the countryside, the young, cosmopolitan Reymund is smitten by Orient's innocent beauty at the same time as she unnerves him with her inhuman aloofness. One day, together with Orient's mother, they venture into the mountains. Deserted by their guide, they are forced to spend the night in the mountains. As morning comes, Reymund proposes to Orient, who firmly rejects him. Nevertheless, before Reymund departs to his business routine in the city, the two enter into a mutual agreement that he will visit Orient in her home every Saturday until his death. Just when Orient realizes her emotional attachment to Reymund, he meets with a fatal accident on his horse. He pays a final ghostly visitation, calling Orient to his deathbed where, at long last, she confesses her love while he exhales his last breath.

On the surface, Spofford unfolds Reymund's and Orient's star-crossed love with the help of stereotypical opposites of white, middle-class female subordination and male dominance, set in a morally superior countryside as opposed to a materialist urban sphere. As "impersonation of morning *and* the East,"[4] the figure of Orient adds racial and imperial dimensions to those of sexual and natural dominion, all of which constitute recurring thematic concerns in Spofford's writings.[5] In Orient, Spofford conjoins the characteristics of Eos and Vespasia from "Desert Sands" (*The Cosmopolitan Art Journal*, December 1860), who embody the dualities of angel/demon and West/East as enforced by Western patriarchal imperialism.[6] Like Eos, Orient is a "sacred" being, who resides in a sphere "as pure and innocent as Eden" apart from the "trifle and artifice and the hollow way of the world."[7] Like Vespasia, she possesses a fearless independence, which causes Reymund to "shudder," as if he "had pledged a bridal ring upon the finger of a ghost."[8] The weird, then, lies not alone in Reymund's ghostly apparition to Orient before his death, which, as Jennifer Patricia Bann has pointed out, lends "the supernatural sentimental rather than terrifying effect."[9] As I want to suggest, the weird also relates to Reymund's increasing ontological uncertainty, which hinges on Orient's "half-fantastic"[10] Otherness, which opens up new spaces and ways of being.

Notably, events in "D'Outre Mort" are structured spatially. Alongside the opposition between country and city, three spaces shape the events: the mountains, Orient's garden, and the home of Orient and her mother. This arrangement seems to imply a linear development from "wilderness" to pastoral and domesticity. However, Orient's presence interconnects these spaces like a triptych and incorporates what Paul Ricoeur has called an episodic chronological and a configurational nonchronological temporal dimension, both of which involve the experience of time as historical and finite.[11] Moreover, the mountains and the garden with their more-than-human agencies are associated with primordial deep time. Time and space, then, together with the mythologizing of Orient, bring on Reymund's existential crisis.

The vertical thrust of the mountains in "D'Outre Mort" is still partly rooted in romantic notions of idealism and transcendence based on an understanding of the correspondence between nature and spirit. The mountains and Orient are practically interchangeable in their awe-inspiring beauty and spiritual sublimity. Whenever Spofford has us look at Orient through Reymund's desirous gaze in particular, his focalization conflates women and nature in what Annette Kolodny in *The Lay of the Land: Metaphor as Experience and History in American Life and Letters* (1975) has identified as land-as-woman metaphor, so that the imperial-territorial dimensions of the story become obvious. Yet, as the heterodiegetic narrative voice makes clear, Orient perturbs Reymund with an irrational fear. Intriguingly, Spofford phrases this fear in terms of a holy as well as a cosmic terror: Reymund not only "crosse[s] himself in his heart" when approaching her, he also observes her "as one watches a clear planet glow steadily from the soft, golden sky."[12] To him, Orient's "intangible" nature equals that of the "indifferent" mountains and invokes the eerie feeling as if his "beautiful mistress was after all a ghoul."[13]

Spofford further endows Orient's disconcerting nature with mythological qualities, drawing on images from ancient, pagan myth. According to German philosopher Hans Blumenberg, it is through myths and their continuous re-narrations that humans attempt to reduce "the absolutism of reality," that is, the human lack of controlling "the conditions of [their] existence."[14] Through myths, he expounds, unknown and overpowering dangers lurking beyond the horizon can be anticipated, while an "'existential anxiety' (*Lebensangst*)" may be rationalized into fear.[15] In this way, the mythological universe provides an artistic and aesthetic "shield" against the terrors of an unmanageable environment in order to contain some undefined monstrosity so that humankind can "be at home in the world."[16] Despite the apotropaic function of myths, they "concentrate" and aesthetically enhance the horrors they try to keep at bay by giving them a poetic gestalt.[17] In myths, residues of an archaic terror remain latent, while their artistic representations convert "something unnerving" into something manageable, or at least accessible.[18] Hence, when Spofford consistently refers to Orient as the embodiment of dawn and the ancient goddess of Eos, she at once attenuates and aesthetically intensifies the terrors regarding the lack of control over the conditions of human existence.

In Orient, Spofford combines the beauty of the goddess Eos with her titanic procreative and cosmological powers.[19] Where Reymund perceives adversity and indifference in nonhuman nature, Orient embraces living "beings," who invite her and talk to her "of the beginning of the earth" and "the end of things."[20] Orient thus shatters Reymund's understanding of the mountains as inanimate backdrop to human actions, as well as his self-orientation in the short-lived, historical time of humankind—which were common nineteenth-century notions that foregrounded historical progress and relegated "nature" to the background as an immobile fixture.[21] Furthermore, Orient's being invokes a world that is, in itself, finite and subject to change and, hence, an agent at once powerful and vulnerable rather than, as Reymund insinuates, a hostile opponent that needs to be subjugated. Indeed, Reymund's name, which means both "protection" and "male rule,"[22] reflects his vacillating between protecting Orient and dominating, even annihilating, any threats to human life by greater, unknown forces—including Orient. Evidently, Reymund and Orient not only hold disparate views regarding their environment, but they also inhabit different worlds.[23]

Not surprising, it is Orient, who, despite the "downward terrors" of their descent and the abandonment of their guide, delights in "the wild beauty of the thing,"[24] when they seek shelter for the night on a rocky ledge, where they witness a spectacular sunset:

> This valley, filled with rolling vapour, whose volumes, smitten by sunset, were fused in splendid colour, made a pavilion of cloud beneath them where billows of fleecy crimson and shining scarlet curdled together intro creamy crests, here seeming to lash in feather-white foam against the base of some crag, and there letting a late sunbeam plough through spaces of a violet-dark drift till they were all inwrought with gold. Above them the cold and mighty heaven was already faintly but thickly strewn with stars.[25]

Poised on the edge of the abyss, the travelers behold the spectacle of a nondescript, luminous mass of light, air, and clouds, swirling and clashing into each other. The fantastic vision resembles the diffuse shapes of galactic or planetary nebulae known only from a vast distance.

Even though Spofford may have been familiar with the nebulae discovered prior to and throughout the nineteenth century, she was not at all interested in joining the ranks of the realists, whose emergence after the Civil War signaled a literary intent upon verisimilitude and "truthful" imitation. Instead, she self-confidently pursued what Dorri Beam has termed a "highly wrought" style.[26] Taking up the cudgels for women's artful yet extravagant writing (as opposed to women's maudlin sentimental fiction), Beam places Spofford alongside such writers as Fuller or Wharton, whose "particular aesthetic and feminist rationale" aimed to "generate alternative models of gendered self and desire," probe into politically more radical themes, and test their achievements against those of their male peers.[27] As Beam shows, Spofford maintained "a romantic insistence on expression"— which impressed contemporaries and later critics alike with its dazzling imagination and baroque expression—but also adopted an aesthetic with which to free the "inherent spirit, inherent function, and inherent beauty" of things, and which substituted transcendence with an "earthbound relentlessness."[28]

The eerie, nebula-like spectacle in "D'Outre Mort" testifies to the materialist agencies and alternative ontologies that Beam points out in Spofford's ornamental aesthetic. With the travelers positioned in between the fantastic sunset below and the cold night sky above, the scene also constitutes what Dana Luciano has identified as a "radical decentering of the human,"[29] to which only Orient—the daughter of the Titans—responds with delight. In contrast, her mother feels transported to the sphere of the dead; and Reymund drinks a liqueur distilled of "oily sunshine,"[30] as if to ritually imbibe the more-than-human forces, which determine over human life and death, albeit to no avail. Throughout the night, undefined apparitions visit yet elude his intelligibility: "some uncreated thing, some phantasm of his brain, or that of some celestial being, some resident of vast spaces, or only a wild beast, a big, brown bear, roving on their tracks and coming to peer about their unprotected bivouac."[31] When the travelers descend from the (planetary) heights and time into the (earthly) valley with its finite temporality, they have undergone important changes. Orient returns with an awakened wonderment about the world around her, preferring "natural religion before theology" and "natural history ... before the petty struggles of warriors."[32] Her mother feels happily restored to her domestic comforts. Only Reymund is profoundly shaken in his self-orientation in space and time.

His ontological confusion continues when he visits Orient's garden, even though it is a semi-cultivated space unlike the mountain heights. Still, he finds himself confronted by weird, unknown exterior forces and agencies:

> ... vine and tendril, leaf, and spray, and branch, and blossom, all wrought themselves to a delicious tangle of perfume, and rustle, and colour. ... a gladiolus reared his flames, a larkspur still blew out a perfect fragrance, while mourning-brides, and gillyflowers, and spiked lavender, and pansies sowed the air with their old-fashioned sweetness. The soft, lonely sky stretched away over the garden and the meadows to haze itself round low and distant woods, and all the empty air seemed sad and desolate between – the fullness and richness of life at its high noon touching close upon the anti-climax of desert solitude. ... These lichen-covered apple trees had shed ... the rosy snow of their petals around [Orient's] head; these gnarled old bergamots had dropped their pulpy globes into her hands ....[33]

Orient's garden has a profuse life of its own, combining an Orientalist eroticism with a decidedly material autonomy, at once alluring and self-absorbed. While it seems to be there for Reymund's taking, his presence actually does nothing to animate this enigmatic paradise. Rather, the motions, scents, and colors of its flowers cast an overwhelming spell on *him*, defiant of as well as independent from his presence.

As Jeffrey Andrew Weinstock has remarked, it is tempting "to feminize"[34] Spofford's fantastic landscapes with their gemstones and flowers. For nineteenth-century feminists and nonfeminists alike, flower language was a popular means of un/veiling and exploring feminine sexuality, including its embodied and/or immaterial qualities. To Reymund, the garden is nothing but "the shadow of Orient herself, reduced to dumb and to material things."[35] It is certainly true that the flowers and plants do not *speak* (as opposed to, for instance, the flowers in the theosophical essays of Spofford's transcendentalist predecessor Margaret Fuller). However, their material qualities—scents, colors, movements—"speak" volumes regarding their being in a world that is different from the world inhabited by Reymund. It is a world, which turns away from transcendence and—oh, the horror!—toward a deeply sensuous, embodied self-possession and autopoiesis, in the sense of self-making.

The story does not end here, however. To combat his existential crisis, Reymund steps up his efforts to win and possess Orient. As he does so, he begins to transform into an "apparition,"[36] which, ironically, highlights his human flaws and mortality, so that his specter of male dominion blurs with that of empire and chauvinism within a historical, finite temporality. At this point, Spofford clearly overdraws the contrast between Orient as female, transcendental, pastoral Other versus Reymund as male, profane, urban self, thus accelerating Reymund's inevitable death and, with it, its psychological terrors: "But alone we come into this world, alone we go out of it. Neither Orient nor another could, for all eternity, give the tone to any soul; that discord or harmony which one shall make must be the result of one's own being."[37] Notably, any hope for consolation or redemption is located in "one's own being." Given Reymund's inability to attain harmony without Orient by his side, the prospect of discord in the afterlife hardly alleviates death's terrors. Indeed, Reymund's spectral visitation of Orient underscores the irredeemable absence of transcendental comforts when his "look of utter melancholy" and "terrible voice … of death, the tomb, and all corruption" foreshadow that, with his death, he will be "gone beyond recall"[38]—even for Orient.

"The Black Bess": Night Train Coming

With its interconnection of sexuality, technology, and madness, the "The Black Bess" engages with the distasteful aspects of the weird more viscerally than "D'Outre Mort." However, both stories feature strong, independent, and sensuous female characters, who evoke male fears of death. In "The Black Bess" in particular, the homodiegetic narrator's obsession with his fiancée Margaret's face—as opposed to her entire body—can be read as either reflecting his sexual loathing and violent rape fantasies, or his traumatic reliving of a fatal accident.

A self-made parvenu, the nameless narrator reflects about his former days as conductor of a night train of the Great Interior Railway. His memories are defined in particular by an accident, during which the onrushing train—the Black Bess—kills the driver of a hay wagon. After partly recovering from his own injuries and depression, the narrator attempts to return to his work, but is haunted by repeated apparitions of his fiancée's face on the train tracks, which one night culminate in the narrator's illusion of "crash[ing] over and through that one dear thing of all [his] heart."[39] After another attempt at recovery, the hauntings continue. Meanwhile, he puts off the planned marriage and seeks help from Dr. Blanchard but is unable to follow the recommendation of confronting the ever-increasing horrors of the "visitations" of Margaret's face. Finally, he works up the nerve to follow doctor's orders and runs the Black Bess over what he considers to be yet another hallucination, only to discover that it was Margaret's body after all. The incident drives him insane and he finds himself in a mental asylum, where his fiancée's face continues to haunt him. According to the narrator, Margaret survives and helps him recover so that, in the end, they marry and he regains control over his life and fortune.

Susan Amper and Birgit Spengler have offered astute interpretations of "The Black Bess." Amper's psychoanalytical reading illuminates the suppressed sexual male drives, fears, and violent fantasies revealed in Spofford's ghost story,[40] whereas Spengler examines the underlying scopic relations, demonstrating that it is in fact the narrator, who becomes an object of the "libidinous looks"[41] of the female gaze, which threatens his sense of masculine domination and results in his hysteria. The focus of my examination lies on the spatial underpinnings of Spofford's story, which show the narrator experiencing space and time as transformed by technological processes—here the mechanical locomotion of the Black Bess. It is an encounter with the weird, in that new human and nonhuman worlds reveal themselves to the narrator, which expand, collide, and enmesh with those he knows and understands a t he same time as they wrest the control over his life from him.

Framed by the external focalization of the narrator, who looks back on the period of his life prior to his social ascent from rags to riches, events are structured as a succession of train rides. These night "runs" on the narrator's "iron steed" are punctuated, first, by the catastrophic accident with the teamster of the hay wagon and, later, the ghastly apparitions of Margaret's face, as well as several sojourns for recovery in domestic spaces and the "whitewashed prison-room" of a mental asylum before his supposed final recovery.[42] Indeed, spatially and temporally, the remembered events appear as one progressive "train" of human technology into unknown spheres and spaces, which at first involve as much exhilaration as they involve terror. Although excited about being the "master" of the Black Bess, who earns "his living by the sweat of [his] brow," the narrator initially dreads "the darkness and the responsibility," and embarks upon his new routine "with a thousand tremors."[43] Yet h e omes to enjoy, even crave, the pleasurable thrills of steering the train into the dizzying heights of remote mountains and the dark hollows of subterranean spheres. However, the accident and the recurring "apparitions" of Margaret's face, intensify the horrors attached to his mastery of the Black Bess, while also rupturing the apparently linear temporal sequence of events.

Significantly, the traversing of time and space with the Black Bess means entering unknown worlds, whose atmospheric and abstract qualities of space and motion are reminiscent of Joseph Mallory William Turner's famous painting *Rain, Steam, Speed—The Great Western Railway* from 1844.[44] The painting dramatizes the collision between the old agricultural social order and the fast, new industrial power of humankind by juxtaposing the clear and solid iron chimney of the onrushing train to the rain- and light-suffused landscape. Here, hard forms and clear colors dissolve in the diffuse atmosphere of the weather and the daylight. In her story, Spofford likewise creates an atmospheric, abstract experience of space. Yet she links it to night and darkness, as well as the Black Bess and her mechanical powers—her "panting" steam, revolving wheels, and glowing headlight, which make her tread "the air" as "light as a feather."[45] Her impressive industrial powers enable the narrator to enter fascinating new worlds in all weathers, be it the tracks' shimmering "ladder of light" on a rainy night, the fleecy, "thick haze" of starbeams in the spring, or the "velvety flakes" which "draped spray and twig of the woodland" under a "grey and dazzling" night sky in winter.[46] As if to reverse Turner's contrast, in Spofford's story, the atmospheric, abstracted spaces of industrial human powers, speed, and darkness, violently clash with the clear forms of the agricultural and natural surroundings of nineteenth-century US America as the Black Bess tears through the rural landscape and rips through the oxen-drawn hay wagon. However, the horror lies less in the violent destruction of a pastoral society by human industrial powers, but rather in the absolute lack of human control, when the "weird distinctness" of the "great steaming oxen" and "the drunken teamster" in the "pillowy masses" of the hay are "scatter[ed] all to flinders" under the "short, sharp shrieks of Black Bess."[47] Evidently, neither the inebriated teamster, nor the taken aback narrator-engineer are masters of their "steeds" or "wheels" and, hence, of the worlds they seek to dominate.

Strikingly, after the accident, the formerly so exhilarating atmospheric experience of novel spaces and modes of travel also spirals out of control. Now, the apparition of Margaret's face becomes an obstacle to motion and progress and superimposes itself on the new spatial visions. At the same time as "the Face" enhances their beauties, it unleashes and amplifies the monstrous powers of the Black Bess that the narrator believes to be able to control.[48] The beloved iron steed now emerges as "belching monster" and, as an extension of the narrator's body, it crushes Margaret "with all the mighty revolutions of my ponderous, red-hot iron" and "all the murderous purpose of this jumping, plunging fiend that I alone controlled, ... sundering socket and mangling flesh."[49] Both Amper and Spengler have pointed out the metaphorical act of rape and mutilation in this passage, in which the brutality of domination merges issues of power, gender, vision, empire, and environment.[50]

Moreover, when taking into consideration the interconnection between the spatial and the industrial-technological dimensions of "The Black Bess," it is important to consider that, as Scott Kirsch reminds us, while technology is a means of "dominating" and "defining nature," there is also a dialectic between society and technology: "Technology, as means ... is always shaped and negotiated by society, while still serving (quite practically) to influence *our* behavior."[51] This means that the Black Bess becomes not only an extension of the narrator, but the narrator's act is also partly produced by the mechanics of his iron horse. In fact, he speculates that the causes for his hallucinations lie in his "ruined" vision, or "that some necessary spring in my mechanism had become loosened."[52] The mechanics of the Black Bess inextricably intermesh with the narrator's human and social nature. So much so, that the narrator senses himself becoming "metallic":

> One day a thundergust has swept across the sky. I have watched its purple masses fold and lap and let their fire down to earth, and, as if I were myself metallic, have felt the electric current coursing down the countless rods that protect my prison-house. A curious sensation has come over me, as if the blood had turned about and were running the right way in my veins, I am conscious of its fresh, free tingling, as if I were just made.[53]

According to the narrator's version of the story, this moment marks a turning point, which returns him to clarity regarding the boundaries of what it means to be human. Assuring himself of the flesh-and-blood humanity of his nurse—he convinces himself that she must be Margaret—by literally probing under her skin with a knife and drawing her blood, he claims to fully regain his sanity (and, by implication, his humanity as well). However, the concluding sentence of the story belies such a sanguine closure. Instead, the narrator's final vow that "the earth shall be ransacked" for his wife and the mother of his children to atone for his earlier sins suggests that, with the onslaught of human technological and industrial powers, the boundaries of being human have significantly shifted.[54] Given the narrator's questionable recovery of his sanity, it portends a shift that involves horrors yet to come.

CONCLUSION: THE WEIRD IN THE INTERSTICE

By focusing on experiences of the weird in two of Spofford's supernatural tales, I have argued that there is a discernible, increasing engagement with more-than-human powers, which prefigures, and even fulfills, Lovecraft's notions of cosmic fear. These insights corroborate Ann and Jeff Vander-Meer's introductory statement in *The Weird: A Compendium of Strange and Dark Stories*, namely that "the Weird often exists in the interstices" and "can occupy different territories simultaneously."[55] Furthermore, in Spofford's short fiction the coexistence of the weird alongside the romantic and the gothic produces fruitful tensions regarding questions of genre and the politics involved in these scholarly formations and practices. Indeed, the study of the weird in Spofford's fantastic short stories is particularly relevant in light of Ellen Ledoux's recent demand for a broader approach to women's gothic fiction. Ledoux calls for a reconsideration of the "Female Gothic" in order to accommodate the diversity and breadth of what she has identified as a too narrow categorization and revisionist ideology, dating back to second-wave feminist scholarship.[56] At the same time, the gender politics of the "Female Gothic" are a useful reminder that, conversely, Lovecraft's cosmic lens may obscure the power relations underlying existential human struggles, even if such matters can indeed pale in view of planetary scales. Hence, examining the weird in the interstices enables constructive inquiry in and across genres as well as critical dialogue about the reevaluation as well as continued relevance of the politics involved in literary criticism.

NOTES

1. H.P. Lovecraft, "Supernatural Horror in Literature," in *Supernatural Horror in Literature and Other Literary Essays*, ed. Darrell Schweitzer (Rockville, MD: Wildside Press, 2008), 19.
2. Lovecraft's notorious racism has been causing considerable controversy among scholars and critics. Some—among them S. T. Joshi—have prioritized Lovecraft's literary and theoretical legacy to his racialist biases. Others—among them Daniel José Older—have argued for making Lovecraft's racism a visible part of his broader legacy. See Carl H. Sederholm and Jeffrey

Andrew Weinstock, "Introduction: Lovecraft Now," *Journal of the Fantastic in the Arts* 26, no. 3 (2015): 445.
3. Lovecraft, "Supernatural Horror in Literature," 173.
4. Harriet Prescott Spofford, "D'Outre Mort," in *The Moonstone Mass and Others*, ed. Jessica Amanda Salmonson (Ashcroft, BC: Ash-Tree Press, 2000), 16 (emphasis added).
5. The perhaps most famous examples of Spofford's gothic stories that voice these concerns are "The Amber Gods" (*Atlantic Monthly*, January and February 1860) and "Her Story" (*Lippincott's*, December 1872).
6. Cynthia Murillo, "The Spirit of Rebellion: The Transformative Power of the Ghostly Double in Gilman, Spofford, and Wharton," *Women's Studies* 42 (2013): 756.
7. Spofford, "D'Outre Mort," 17–18, 29.
8. Ibid., 18–20.
9. Jennifer Patricia Bann, "Spirit Writing: The Influence on Spiritualism on the Victorian Ghost Story" (dissertation, University of Sterling, 2007), 112. In her brief reference to Spofford's story, Bann notes that the literary convention of ghost sightings prior to a loved one's death, "affirm[s] the sacred status of the marital bond."
10. Spofford, "D'Outre Mort," 17.
11. Paul Ricoeur, "Narrative Time," *Critical Inquiry* 7, no. 1 (1980): 178. Addressing literary scholars' neglect of the reciprocity between narration and temporality, Ricoeur argues, among other things, that fictional narratives contain various "degrees of temporal organization" (170). He notes that every narrative contains a dialectic between an "episodic dimension" of successive events (chronological dimension) and a "configurational dimension" (178) of meaningfully arranged patterns of events generated by the plot (nonchronological dimension).
12. Spofford, "D'Outre Mort," 17–18.
13. Ibid., 19–20.
14. Hans Blumenberg, *Work on Myth*, trans. Robert M. Wallace (Cambridge and London: MIT Press, 1985), 3–4.
15. Ibid., 6.
16. Ibid., 14 and 113.
17. Ibid., 14.
18. Ibid., 75.
19. For the attributes and reception of Eos from antiquity to modernity in literature and the arts, see Tobias Leuker, "Eos," in *Brill's New Pauly Supplements I, Volume 4: The Reception of Myth and Mythology*, eds. David van Eijndhoven, Christine Salazar, and Francis G. Gentry (Stuttgart, Germany: Metzler, 2010), http://dx.doi.org/10.1163/2214-8647_bnps4_e330980.
20. Spofford, "D'Outre Mort," 18–19.

21. Robert T. Tally, Jr., *Spatiality* (London and New York: Routledge, 2013), 30–33.
22. See "Ray," "Raymond," and "Reimund" in *A Dictionary of First Names*, eds. Patrick Hanks, Kate Hardcastle, and Flavia Hodges (Oxford and New York: Oxford University Press, 2006), 300, 302.
23. Spofford adapts elements of the medieval romance *Melusine*, according to which Raymondin discovers Melusine's monstrous creatureliness when he breaks his promise to never come and see her on Saturdays.
24. Spofford, "D'Outre Mort," 21–22.
25. Ibid., 23.
26. Dorri Beam, *Style, Gender, and Fantasy in Nineteenth-Century American Women's Writing* (New York: Cambridge University Press, 2010), 1–33.
27. Ibid., 2.
28. Ibid., 157–61.
29. Dana Luciano, "Geological Fantasies, Haunting Anachronies: Eros, Time, and History in Harriet Prescott Spofford's 'The Amber Gods'," *ESQ: A Journal of the American Renaissance* 55, no. 3 (2009): 289.
30. Spofford, "D'Outre Mort," 23.
31. Ibid., 24–25.
32. Ibid., 26.
33. Ibid., 28.
34. Jeffrey Andrew Weinstock, "The Queer Time of Lively Matter: The Polar Erotics of Harriet Prescott Spofford's 'The Moonstone Mass'," *Women's Studies* 46, no. 8 (2017): 12.
35. Spofford, "D'Outre Mort," 28.
36. Ibid., 27.
37. Ibid., 30.
38. Ibid., 32–34.
39. Ibid., 90.
40. Susan Amper, "Male Fears and Fantasies in Harriet Prescott Spofford's 'The Black Bess'," *West Virginia University Philological Papers* 47 (2001): 29–36.
41. Birgit Spengler, *Vision, Gender, and Power in Nineteenth-Century American Women's Writing, 1860–1900* (Heidelberg: Universitätsverlag Winter, 2008), 96.
42. Spofford, "The Black Bess," 83 and 97.
43. Ibid., 83.
44. See the link at the National Gallery, London, https://www.nationalgallery.org.uk/paintings/joseph-mallord-william-turner-rain-steam-and-speed-the-great-western-railway.
45. Spofford, "The Black Bess," 83.
46. Ibid., 86, 92, and 94.
47. Ibid., 84.
48. Ibid., 88.

49. Ibid., 88 and 96.
50. Amper, "Male Fears and Fantasies in Harriet Prescott Spofford's 'The Black Bess'," 29; and Spengler, *Vision, Gender, and Power in Nineteenth-Century American Women's Writing, 1860–1900*, 100–101.
51. Scott Kirsch, "The Incredible Shrinking World? Technology and the Production of Space," *Environment and Planning D: Society and Space* 13 (1995): 535.
52. Spofford, "The Black Bess," 91.
53. Ibid., 97.
54. Ibid., 98.
55. Ann and Jeff VanderMeer, Introduction to *The Weird: A Compendium of Strange and Dark*.
56. Ellen Ledoux, "Was There Ever a 'Female Gothic'?" *Palgrave Communications* 3 (2017): 1–7.

Bibliography

Amper, Susan. "Male Fears and Fantasies in Harriet Prescott Spofford's 'The Black Bess'." *West Virginia Philological Papers* 48 (2001): 29–36.
Bann, Jennifer Patricia. "Spirit Writing: The Influence on Spiritualism on the Victorian Ghost Story." Dissertation, University of Sterling, 2007.
Beam, Dorri. *Style, Gender, and Fantasy in Nineteenth-Century American Women's Writing*. New York: Cambridge University Press, 2010.
Blumenberg, Hans. *Work on Myth*. Translated by Robert M. Wallace. Cambridge and London: MIT Press, 1985.
Hanks, Patrick, Kate Hardcastle, and Flavia Hodges, eds. *A Dictionary of First Names*. Oxford and New York: Oxford University Press, 2006.
Kirsch, Scott. "The Incredible Shrinking World? Technology and the Production of Space." *Environment and Planning D: Society and Space* 13 (1995): 529–55.
Ledoux, Ellen. "Was There Ever a 'Female Gothic'?" *Palgrave Communications* 3 (2017): 1–7.
Leuker, Tobias. "Eos." In *Brill's New Pauly Supplements I, Volume 4: The Reception of Myth and Mythology*, edited by David van Eijndhoven, Christine Salazar, and Francis G. Gentry. Stuttgart, Germany: Metzler, 2010. http://dx.doi.org/10.1163/2214-8647_bnps4_e330980.
Lovecraft, Howard Phillips. "Supernatural Horror in Literature." In *Supernatural Horror in Literature and Other Literary Essays*, edited by Darrell Schweitzer, 13–112. Rockeville, MD: Wildside Press, 2008.
Luciano, Dana. "Geological Fantasies, Haunting Anachronies: Eros, Time, and History in Harriet Prescott Spofford's 'The Amber Gods'." *ESQ: A Journal of the American Renaissance* 55, no. 3 (2009): 269–303.

Murillo, Cynthia. "The Spirit of Rebellion: The Transformative Power of the Ghostly Double in Gilman, Spofford, and Wharton." *Women's Studies* 42 (2013): 755–81.

Ricoeur, Paul. "Narrative Time." *Critical Inquiry* 7, no. 1 (1980): 169–90.

Sederholm, Carl, and Jeffrey Andrew Weinstock. "Introduction: Lovecraft Now." *Journal of the Fantastic in the Arts* 26, no. 3 (2015): 444–49.

Spengler, Birgit. *Vision, Gender, and Power in Nineteenth-Century American Women's Writing, 1860–1900.* Heidelberg: Universitätsverlag Winter, 2008.

Spofford, Harriet Prescott. "D'Outre Mort." In *The Moonstone Mass and Others*, edited by Jessica Amanda Salmonson, 16–34. Ashcroft, BC: Ash-Tree Press, 2000.

———. "The Black Bess." In *The Moonstone Mass and Others*, edited by Jessica Amanda Salmonson, 82–98. Ashcroft, BC: Ash-Tree Press, 2000.

Tally, Robert T., Jr. *Spatiality.* London and New York: Routledge, 2013.

VanderMeer, Ann and Jeff. Introduction to *The Weird: A Compendium of Strange and Dark Stories*, xv–xx. Edited by Ann and Jeff VanderMeer. London: Tor, 2012.

Weinstock, Jeffrey Andrew. "The Queer Time of Lively Matter: The Polar Erotics of Harriet Prescott Spofford's 'The Moonstone Mass'." *Women's Studies* 46, no. 8 (2017): 1–15.

CHAPTER 4

The Weird and the Wild: Media Ecologies of the Outré-Normative

Julius Greve

The oldest and strongest emotion of mankind is fear, and the oldest and strongest kind of fear is fear of the unknown. These facts few psychologists will dispute, and their admitted truth must establish for all time the genuineness and dignity of the weirdly horrible tale as a literary form.

— H. P. Lovecraft, "Supernatural Horror in Literature"[1]

Wildness is not the lack of inscription; it is inscription that seeks not to read or be read but to leave a mark as evidence of absence, loss, and death. Wildness must take us into its mottled embrace and press us to stare into those places of slippage between language and experience and life and death; wildness can give us access to the unknown and the disorderly, and we will enter there at our own risk.

— Jack Halberstam, "Wildness, Loss, Death"[2]

How to think "the weird" or "weirdness" not only alongside the tradition of Gothic literature and culture, and the category of "the uncanny,"

J. Greve (✉)
University of Oldenburg, Oldenburg, Germany

© The Author(s) 2019
J. Greve and F. Zappe (eds.), *Spaces and Fictions of the Weird and the Fantastic*, Geocriticism and Spatial Literary Studies,
https://doi.org/10.1007/978-3-030-28116-8_4

but rather vis-à-vis the notion of "the wild"? "Wildness"—a central concept in American letters at least since Henry David Thoreau's theorizations of it—is fundamentally other to the "cosmic fear"[3] evoked by the genre of Old Weird fiction or weirded storytelling. It is different insofar as it emphasizes a particular ethical relationship between humankind and its surroundings, between the individual organism and the environment, different from the ethical and political implications of H. P. Lovecraft's creatures. In fact, the worship of wildness in the Thoreauvean tradition is in many ways coextensive with the doctrines of deep ecology, of which the Beat poet, mythographer, and ecologist Gary Snyder is one of the main representatives. However, the ways in which the genre of "the weird" has, in recent years, been contextualized and reframed, within the environmental humanities and speculative realist circles, suggest that the discourses of weirdness and wildness could *and should* be put into a productive dialogue.

The two epigraphs are in many ways emblematic of both the similarities and the differences between the weird and the wild. The first is a fairly known passage from Lovecraft's seminal essay "Supernatural Horror in Literature" (1927), describing the core qualities and subject matter of weird fiction, as he sees it. He hones in on the what he calls "the oldest and strongest emotion of mankind"—"fear of the unknown"—and argues that this "truth must establish for all time the genuineness and dignity of the weirdly horrible tale as a literary form."[4] Lovecraft's universalist ontological inference concerning human emotion is thus paired with a statement concerning literary genre conventions. The second quote is a more recent take on the term "wildness" by queer theorist Jack Halberstam, who will be one of the many (and unlikely) conversational partners of Lovecraft's in this chapter. Both paragraphs, arguably, are not merely linked to, but representative of, the two literary and cultural US traditions of the weird and the wild.

Halberstam argues that the notion of "wildness" does not refer to a prelapsarian mode of cultural practice. Rather, it concerns an unreadable type of "inscription" which resists both the mode of decipherability and decipherment, in order "to leave a mark as evidence of absence, loss, and death." This subcultural and, hence, anti-universalist "wild writing," therefore, "can give us access to the unknown and the disorderly," and, as Halberstam notes from his non-heteronormative subject position, "we will enter there at our own risk."[5]

Why should there be a dialogue in need of facilitation, between a racist author's understanding of "genuineness and dignity" vis-à-vis "literary form,"[6] and a contemporary queer theorist's notion of "those places of slippage between language and experience and life and death"?[7] I argue that irrespective of there being multiple, seemingly insurmountable differences between writers like Lovecraft and Halberstam (or other literary and/or critical "pairs" from the long list of authors of the weird and the wild, such as, for instance, Robert W. Chambers and Jack London, or Annie Dillard and Thomas Ligotti), these two traditions have existed alongside each other—the conceit of this chapter being that the historical parallel of both literary-cultural trajectories from the early nineteenth century onward is not coincidental but perhaps even points to the co-constitutive of both weirdness and wildness in American culture. In a sense, the wild and the weird recapitulate the relationship others have described as the difference between "transcendentalism" and "descendentalism," or between the worship of nature and the indifference of the latter.[8] In part following Mark Fisher's taxonomy in *The Weird and the Eerie* (2016),[9] the first or primary tradition I am concerned with is "the weird" and the second is "the wild." And even though the second has a much longer discursive history, it is the notion of "weirdness" as an aesthetic quality, of "weirding" as cultural and literary process, or simply of "*the* weird" as a genre conception that I am particularly interested in. The viewpoint of "weirdness" is that from which I will examine the specifically American casting of "wildness" in a new light, tracing the conceptual affinities that the generic notion of the weird (and its inflections as the Old Weird and the New Weird) entertains in regard to earlier traditions, genres, and discourses in American culture—chief of all, the wild.

I will, thus, attempt a comparison between the weird and the wild, by means of discussing both seminal and slightly more obscure, or marginal works by writers as different as Thoreau, Snyder, Lovecraft, London, Tzvetan Todorov, Jamaica Kincaid, Ligotti, and Halberstam—and what I mean by "seminal" and "marginal" should in each individual case be understood with respect to both the problematics of canonization and of genre discourse. The discussion will be cast in terms of what each and every one of these writers treats of in their work, in various degrees and in multiple ways: the anxieties caused in and by what has been called the Anthropocene—the era in which, as McKenzie Wark puts it in *Molecular Red* (2015), "[t]he human is no longer that figure in the foreground which pursues its self-interest against the background of a wholistic, organicist cycle that the human might perturb but with which it can remain in balance and harmony, in the end, by simply *withdrawing* from certain excesses."[10]

In addition to the literary manifestations of the equally "weird" and "wild" dynamics of American culture, a broader media ecological framework is needed to account for these dynamics. Examples in this regard will include the work of artists as diverse as the filmmakers Werner Herzog, Wu Tsang, John Carpenter, David Lynch, and Nic Pizzolatto. Finally, I will argue that what William Cronon called "the trouble with wilderness,"[11] in relation to the weird and the wild in American media ecologies, fundamentally leads to the question concerning ethico-aesthetic forms of being and expression. In other words, it leads to a genre-specific take on the issue of normativity in the context of artistic practices, or what I will call the "outré-normative."

"A Sense of Dread" and Melting Icecaps

Starting with the conception of the weird, the etymology of which includes the meaning of oddness or strangeness, the Old English usage of the term as a noun (*wyrd*), denoting predetermination or fate, or simply "becoming,"[12] the present literary context is the genre which Lovecraft came to define as a specific literary mode that had its place in the direct vicinity of the uncanny, the marvelous, the Gothic, and the fantastic—the latter of which is almost coextensive with the literary and aesthetic category championed by the notorious author from Providence, Rhode Island. As Lovecraft writes in "Supernatural Horror in Literature," establishing a connection between his own writing and certain forms of British and American Gothic fiction, and, eventually, departing from it: "The true weird tale has something more than secret murder, bloody bones, or a sheeted form clanking chains according to rule. A certain atmosphere of breathless and unexplainable dread of outer, unknown forces must be present; and there must be a hint, expressed with a seriousness and portentousness becoming its subject, of that most terrible conception of the human brain ..."[13] And, as he defines the criteria for judging whether or not a "weird tale" is successful as such:

> The one test of the really weird is simply this—whether or not there be excited in the reader a profound sense of dread, and of contact with unknown spheres and powers; a subtle attitude of awed listening, as if for the beating of black wings or the scratching of outside shapes and entities on the known universe's utmost rim. And of course, the more completely and unifiedly a story conveys this atmosphere, the better it is as a work of art in the given medium.[14]

What can be witnessed in this theorization of weird narration is a refusal of any type of formulaic emphasis on gore, as well as the traditional form of the ghost story, for instance (even though Lovecraft's oeuvre does include ghosts and related creatures). However, he is chiefly concerned with the weird tale being an expression of the human feelings or affects of fear, anxiety, or what he also calls a "sense of dread"—of a malignant "atmosphere" that puts to question the only ostensibly fixed laws of nature upon which our daily cognitive operations as subjects rely. If the notion of the post-Freudian "uncanny," as conceived of by the literary theorist Tzvetan Todorov, was an occasion of a merely temporary suspension of those laws of nature, only to be recuperated at the ending of a given story in favor of rational explanations, "the marvelous" would leave the suspension of the natural order of things as the new status quo, thus resulting in tales at the ending of which gravity, for example, would be defied, leading to the respective protagonists having a new sense of being in the world—irrespective of any resulting "sense of dread" on their part whatsoever. What Todorov named "the fantastic," however, would be that in-between state, from a narratological point of view; that is, a state in between rationalist recuperation and irrationalist defeat of the workings and rules of natural dynamics, as perceived and agreed upon by the scientific and public community at a given point in time. Much like Lovecraft's weird, Todorov's conception of the fantastic has to do with the ambiguity in a given narrative, and the subsequent hesitation induced in its readers or characters "who must decide whether or not what they perceive derives from 'reality' as it exists in the common opinion."[15]

In any event, both in Lovecraft and—to a certain extent—in Todorov, the weird and/or the fantastic[16] modes of storytelling concern a psychological unsettling, which is also a material unsettling both for characters in and readers of tales encompassed by these genres. This is so simply because in terms of the Freudian iceberg-model, which had cast the unconscious as an overwhelmingly large yet hidden portion of the human psyche, the Lovecraftian proposal of a genre founded upon the continuance of an aestheticization of fear would point to the inexorable melting of those icecaps and its below-sea-level portions. In fact, this is one of the ways in which the notion of the weird meets that of the wild in an as yet unexplored fashion—namely, in the context of an era in which humanity becomes a force as devastating as any geological one before it.

An Ethics "Without Fear of Wild Beasts"

For some geologists and theorists of the Anthropocene, the decade of the 1850s is contemporaneous with the early phase of this epoch, in which humanity develops into that globally efficient force that would forever change and irreversibly damage the ecosystem more than any other part of the latter. This coincides with Henry David Thoreau's transcendentalist elaborations of his love for walking in his New England environs—which is, in fact, above all his account of what he terms "the wild" or "wildness." However, for an American thinker such as Thoreau, the concept of wild nature did not include the same bleak outlook as our contemporary theorists and climate scientists do. Far from it: his conception was, arguably, still naively complicit with certain imperialist aspirations of the American people and its myth of a certain destiny—and it would be almost a tautology to say that such a manifest destiny was a weird one, given that one of the meanings of the weird *is* destiny or fate itself.[17]

The wilderness in North America, which is the space for the wild, the material condition or as disposition of the latter, on Thoreau's account, seems to be entirely benevolent toward humans, a vital source for humankind's perseverance as species, benevolent even in its harsh reality. Thoreau states: "We are told that within three miles of the center of the East-Indian city of Singapore, some of the inhabitants are annually carried off by tigers; but the traveler can lie down in the woods at night almost anywhere in North America without fear of wild beasts."[18] Like his friend and fellow transcendentalist Ralph Waldo Emerson, Thoreau only seemingly avoids the cultural context of his Anglo-American forebears from Europe and instead repeats the ideology of the West as the place of promise and hope. Christening the West as a new Eden to come, Thoreau writes some of his most famous lines:

> The West of which I speak is but another name for the Wild; and what I have been preparing to say is, that in Wildness is the preservation of the World. Every tree sends its fibers forth in search of the Wild. The cities import it at any price. Men plow and sail for it. From the forest and wilderness come the tonics and barks which brace mankind. Our ancestors were savages. The story of Romulus and Remus being suckled by a wolf is not a meaningless fable. The founders of every state which has risen to eminence have drawn their nourishment and vigor from a similar wild source. ... Give me a wildness whose glance no civilization can endure.[19]

Others have read this passage in multiple ways and it is certainly up for discussion if Thoreau was either apologetic or mainly critical of the harvesting of natural resources by the processes of civilization and modernization. His account is ambiguous, given that he both praises the turn toward Western lands and seeks to speak up for the preservation of the world.[20] What I am interested in the context of this essay, however, is the ambivalence with which "the wild" or "wildness" comes to be defined in "Walking"—that is to say, if it denotes above all an entirely benevolent force of nonhuman activity or if it is rather a stand-in for all that is unruly and ungovernable in the world. A third option would certainly be "wildness" understood as a disposition or quality that human beings would interpret as mostly or entirely benevolent, but that would eventually turn out to be actually fierce and fatal. Werner Herzog's fascinating documentary *Grizzly Man* (2005) exemplifies this third option. The movie recounts the life and death of the eco- and animal rights activist Timothy Treadwell by means of found footage and interview pieces featuring Treadwell's loved ones. In one of the interviews, the so-called "Bear Biologist" Larry Van Daele says: "If you spend a lot time with bears day after day, there is a calling, that makes you wanna come in and spend more time in the world." This notion of a specific "calling," gives me opportunity to turn to yet another seminal author who, in the early twentieth century, would publish the naturalist novel *The Call of the Wild* (1903), which chronicles the process of uncivilization that a domestic dog undergoes. In the end, after his ordeal of enduring what London terms "the law of club and fang," Buck learns the ways of the wolves in the Yukon region of Canada during the turn of the century Gold Rush, following and resonating with "the sounding of the call," "looking for it as though it were a tangible thing, barking softly or defiantly, as the mood might dictate," and eventually becoming "a long-drawn howl, like, yet unlike, any noise made by husky dog."[21]

While this aspect of the musical unconscious of wildness is certainly crucial—to the degree that it has purchase on the trajectory of the genre conceptions of the weird to begin with—it is in Gary Snyder's work and his theorizations of wildness that we find one of the most defining instances in the history of the concept ever since Thoreau's essay "Walking." In *The Practice of the Wild* (first published in 1990), Snyder suggests that there is an irreducibly ethico-aesthetic aspect in any notion of the wild, by asking: "Where do we start to resolve the dichotomy of the civilized and the wild?"[22] This dichotomy might sound like the often-evoked opposition of internal organization and externally forced domesticity (as can be witnessed in London's tale of the dog Buck in *The Call of the Wild*), yet from Snyder's ecocentric point of view, "Nature is not a place to visit, it is *home*—and within that home territory there are more familiar and less familiar places."[23] Consequently, according to Snyder, the saying that "[t]he world is watching" means that "one cannot walk through a meadow or forest without a ripple of report spreading out from one's passage."[24] What this entails for the concept of wildness or the wild is that there is a certain type of immanent communication, social order, or resonance that is different from that of a so-called "order of civilization" but that is itself, nonetheless, capable of intelligence and sociality—a type "report" not unlike the "strange sentience"[25] that Kate Marshall traces in both nineteenth- and twentieth-century examples of weird fiction and new materialist discourse.

Snyder's reasoning, then, results both in "seeing the human and animal as in the same realm," and in the notion that regarding the pressing issue of climate change, the space of "Wilderness" or self-sufficient ecosystems, "may temporarily dwindle, but wildness won't go away."[26] It is striking, moreover, that Synder starts his etymological scrutiny of the concept of the wild with a metaphorical treatment of that word. In Synder's hands, the concept itself becomes animal: "The word wild is like a gray fox trotting off through the forest, ducking behind bushes, going in and out of sight. Up close, first glance, it is 'wild'—then farther into the woods" it recedes to the meaning of "Latin *ferus* (feral, fierce), which swings around to Thoreau's 'awful ferity' shared by virtuous people and lovers."[27]

WEIRD, WILD, WRONG

What is the connection we can draw between the weird and the wild? What might be the similarities and differences between "the sense of *wrongness* associated with the weird—the conviction that *this does not belong*,"[28] as Fisher puts it, and the "unrestrained, insubordinate, licentious," the "unruly," "outrageous," and "ecstatic," as Snyder catalogues some of the dominant ascriptions to the wild, as a concept?[29] There are several points I have raised, so far, leading me to the preliminary conclusion that while the etymologies of the weird and the wild are not isomorphic or coterminous, there is a certain semantic affinity that is unmistakable: with the weird pointing to states of becoming, strangeness, and non-belonging, and the wild referring to the primary and undomesticated disposition of living entities, such as animals, plants, but also human beings, there is an apparent questioning of established *norms* in a given human society suggested by these concepts. This suspension of the normative might very well entail Lovecraft's "atmosphere of breathless and unexplainable dread,"[30] in terms of the disentangling of rigidity and order, for a given set or structure of the social fabric, but it might also lead to an enlarged concept of communication, of being together, and being in the world, suggested by Thoreau's and Snyder's ethics.

Along these lines, Jack Halberstam has, together with a group of colleagues, including Tavia Nyong'o and the late José Esteban Muñoz, revitalized the notion of the wild in the contexts of gender and sexuality studies, queer theory, and critical race studies. Interestingly, Halberstam conceives of what he calls "other regimes of knowing and being that form part of the making and unmaking of modern racialized subjectivity." Further, a queer studies perspective in our context, "a queer inquisition into 'wildness'" would cast the latter "as the space that colonialism constructs, marks, and disavows, as well as a space of vibrancy that limns all attempts to demarcate subject from object, and a space of normativity that holds the deviant and the monstrous decisively at bay."[31] While movies such as John Carpenter's *The Thing* (1982) or David Lynch's *Inland Empire* (2006) have already been linked to the genre of the weird, it is the 2011 documentary fiction *Wildness* by performance artist and MacArthur Fellow Wu Tsang that Halberstam turns to in order delineate queer spaces of the wild—the movie being about a Los Angeles latino/a trans bar called Silver Platter, the story of which is narrated from the perspective or, in fact, "strange sentience"[32] of the place itself, and the subcultures and scenes it houses or encompasses.[33] Yet irrespective of the two conceptual and cultural traditions of the weird and the wild: If weirdness is another word for wrongness and if wildness is another word for unruliness, then, both are oppositional to the structure and proliferation of hetero-normativity and to the norm of imperialism and its colonies. In our present context, it is no coincidence that it is the social structure of the family unit and its *oikos* (or household) that is negotiated by many contemporary works of American weird fiction, as different as Jamaica Kincaid's short story "My Mother" (1978), Joyce Carol Oates "Family" (1989), and Thomas Ligotti's "The Frolic" (1989).

Consider Kincaid: In the late seventies, the Caribbean-American writer who would eventually come to be known for her autobiographical fiction—such as *Lucy* (1990) or *Mr. Potter* (2002)—published a short story called "My Mother," which, in a highly poetic style, depicts the dynamic tensions produced both by the relation between a mother and her daughter and the postcolonial environment of the Carribbean, both being in a constant state of becoming. And, it has to be remembered, these are tensions that are true not only to the genre but to the etymology of the weird. In the following passage, the narrator tells us:

> My mother and I were standing on the seabed side by side, my arms laced loosely around her waist, my head resting securely on her shoulder as if I needed the support. ... I was no longer a child but I was not yet a woman. My skin had just blackened and cracked and fallen away and my new impregnable carapace had taken full hold. My nose had flattened; my hair curled in and stood out straight from my head simultaneously; my many rows of teeth in their retractable trays were in place.[34]

It is the animality that is wedded to the material becoming of the black feminine body that Kincaid depicts in her short story. Yet it is not only a story about the strangeness associated with the liminal states of adolescence. "My Mother" is also concerned with that generic formula of "the call," or strange and unruly noises that pertain both to the weird and the wild, as it is with the notion of decay as a subversive force—a trope common to most types of horror literature. "My mother reached out to pass a hand over my head, a pacifying gesture, but I laughed and, with great agility, stepped aside. I let out a horrible roar, then a self-pitying whine. I had grown big, but my mother was bigger, and that would always be so. We walked to the Garden of Fruits and ... departed through the southwesterly gate, leaving as always, in our trail, small colonies of worms."[35]

While Oates' 1989 science fiction narrative called "Family" revisits the abject terror of decay and what in film studies is called body horror, it also includes the figure of a strong matriarch simply named "Mother" with a capital "M" whose strong regimen in a family stricken with poverty is enforced by a "voice so hard, harsh, brassy, and penetrating, a shout from her had the power to paralyze any of us where we stood,"[36] as the narrator tells us via the focalizing perspective of one of Mother's children. And it is this mention of the voice as the carrier of both a weird "sense of dread" and a wild calling or shouting within that social structure denoted by "the sacred word *family*"[37] that brings me, finally, to the work of Ligotti, who in 2014 was mentioned by Nic Pizzolatto, the creator of the HBO show *True Detective*, as one of his main literary influences (a show whose other very overt referencing of weird fiction consists in the mystique suggested by the word "Carcosa" throughout Season 1; a word that originates in the work of Robert W. Chambers' 1895 tale *The King in Yellow*).[38]

In the same year as Oates' narrative, Ligotti published a tale called "The Frolic." This short story depicts, in a hilariously laconic manner, the evening conversation between a husband and his wife in the small town called Nolgate. This town, as we get to know in the very first sentence of the narrative is also the "site of the state prison,"[39] at which Dr. David Munck works as a psychiatrist. In the course of their conversation, Munck and his wife Leslie consider the idea of actually leaving Nolgate for good, because a certain inmate David works with, called John Doe, has been acting up; in the form of implicitly suggesting that he would try and actually kidnap the Dr.'s daughter Norleen, even though he could not have known about her existence. At the ending the story the terrified parents find out that their daughter is no longer in her room, while a draft from an opened window can be felt throughout their home. This is, then, how Ligotti describes the affective state of the father, as he realizes his daughter has actually been kidnapped:

> 'David?' he heard his wife's voice inquire from the bottom of the stairs. 'Is everything alright?'
> Then the beautiful house was no longer quiet, for there rang a bright freezing scream of laughter, the perfect sound to accompany a passing anecdote of some obscure hell.[40]

It would not be adequate to simply conclude by saying that both the weird and the wild, each in their distinct ways, give rise to reflections, subversions, and critiques of the social realm—a realm grounded on the watershed moment of a lapse from an ostensibly preestablished harmony with nature toward perpetual alienation from and by the latter. While it is true that what turns out to be crucial for both is an emphasis on the ecological and material base of that realm and its experience—along the lines, perhaps, of what Georges Bataille's once termed "base materialism"[41]—as well as of the imagination by which that realm's subversion might be effectuated, there is a particular dilemma that is at the heart of both concepts and their chiasmic interrelations; that is to say, at the heart of the weirdness of the wild and the wildness of the weird. This is the dilemma of the simultaneous de- and re-instantiations of normative structure and dynamics, when it comes to processes of "civilization" vs. "uncivilization"[42] or hetero-normativity and what has been termed its "queering"—or simply any scenario in which literary and social forces appear to be in the act or circumstance of "bending." This dilemma, therefore, I call "the outré-normative," since it both accounts for the external, and externalized, cosmic fear of Lovecraft's universe, and the form of wildness that Halberstam is interested in, namely, a certain animistic "extravagance" in queer and drag-culture, in his examples, which, as he states, "can give us access to the unknown and the disorderly, and [which] we will enter ... at our own risk."[43] Moreover, as the *OED* tells us (admittedly, one of the arch-apparatuses of "inscription" that, in Halberstam's phrasing, "seek ... to read or be read"), the word "outré" signifies that which is "[b]eyond the bounds of what is usual or considered correct and proper; unusual, peculiar; eccentric, unorthodox; extreme."[44] As Edgar Allan Poe (one of Lovecraft's most significant precursors), lets his detective Auguste Dupin reason in the genre defining short story "The Murders in the Rue Morgue" (1841): the murder at hand, which Dupin and the narrator have read about in the newspaper *Gazette des Tribuneaux* at the beginning of the story, and the crime scene of which both of them have subsequently inspected because of its "unusual" circumstance, is "considered insoluble" by the *Gazette* "for the very reason which could cause it to be regarded as easy of solution—I mean for the *outré* character of its features. The police are confounded by the seeming absence of motive—not for the murder itself—but for the atrocity of the murder."[45] It is this very emphasis on form—"absence of motive... for the atrocity of the murder"—rather than content—"the murder itself"—that already shows the disposition of an outré-normative aesthetics, grounded in both the fear and refusal of what is deemed the horrifically senseless or "MALIGNANTLY USELESS world"[46] of the social realm—from the point of view of weirdness, wildness, and wrongness, that is.

Thus, in each of these examples—the *OED*, Edgar Allan Poe's detective fiction, and the present context of Lovecraft's and Halberstam's encounter—the notion of "outré" or "*the* outré" actually resonates with many etymological references I elaborated earlier vis-à-vis the weird and the wild; namely, the oddness or strangeness of the former, and the unrestrained and insubordinate nature of the latter—in other words, the contagious media ecologies of ferality, which may be glimpsed, among other things in Carpenter's movie *The Thing*, whose outré monstrosity in turn uses human and nonhuman bodies as media and puts them to use in "peculiar; eccentric; unorthodox; [and thus] extreme" fashion.

Finally, apart from the visual, textual, and social dimensions of weirdness, wildness, and their external-externalized sense of an "outré" normativity, what the foregoing has demonstrated, in terms of the respective concepts and practices mentioned, from London's *The Call of the Wild* to what might be termed Ligotti's "call of the weird," (which equally reverberates through a plethora of weird body horror masterpieces, including *The Thing*) the literary and cultural expression of the *sonic* character of the outré appears to be key in understanding the relationship of weirdness and wildness in the contexts of genre discourse, media ecology, and the environmental humanities. That is to say, the outré-normative may be a realm whose affects are conveyed in a predominantly sonic fashion, rather than in terms of visually or haptically mediated horror. In any case, it would have to be an essentially *improper* sound that would delineate the normativity of the outré, one that is are distressing as Buck's ferality combined with the "bright freezing scream of laughter, the perfect sound to accompany a passing anecdote of some obscure hell,"[47] all of which being distorted by the effect of the infamous "mad word ...'*Tekeli-li! Tekeli-li!*,'"[48] voiced incessantly by the character Danforth at the ending of Lovecraft's *At the Mountains of Madness* (1936). Put simply, such an inadequate, extravagant, frightful, and extreme sound of the outré-normative would have to be located, that is, between the weird and the wild.

NOTES

1. H. P. Lovecraft, "Supernatural Horror in Literature," in *At the Mountains of Madness: The Definitive Edition* (New York: The Modern Library, 2005), 105.
2. Jack Halberstam, "Wildness, Loss, Death," *Social Text* 32, no. 4 (2014): 147.
3. Lovecraft, "Supernatural Horror in Literature," 107.
4. Ibid., 105.
5. Halberstam, "Wildness, Loss, Death," 147.
6. Lovecraft, "Supernatural Horror in Literature," 105.
7. Halberstam, "Wildness, Loss, Death," 147.
8. See, for instance, Ted Billy's "Descendentalism and the Dark Romantics: Poe, Hawthorne, Melville, and the Subversion of American Transcendentalism," in *A Companion to American Gothic*, ed. Charles L. Crow (Oxford, UK: Wiley, 2013), 151–163.
9. Mark Fisher, *The Weird and the Eerie* (London: Repeater Books, 2016).
10. McKenzie Wark, *Molecular Red: Theory for the Anthropocene* (London: Verso, 2015), xii.
11. William Cronon, "The Trouble with Wilderness: Or, Getting Back to the Wrong Nature," *Environmental History* 1, no. 1 (1996): 7–28.

12. See Jeffrey A. Weinstock, "The New Weird," in *New Directions in Popular Fiction: Genre, Distribution, Reproduction*, ed. Ken Gelder (London: Palgrave Macmillan, 2016), 177, for some of the etymological references.

13. Lovecraft, "Supernatural Horror in Literature," 107.

14. Ibid., 108.

15. Tzvetan Todorov, *The Fantastic: A Structural Approach to a Literary Genre*, trans. Richard Howard (1973; New York: Cornell Paperbacks, 1975), 41.

16. See Todd Spaulding, "The Emerge(d)nt Weird Tale: A Genre Study," *Studies in the Fantastic* 3 (2015/2016): 80, 84.

17. See also on this point and the idea of a naturalist and realist form of weird fiction: Kate Marshall, "The Old Weird," *Modernism/modernity* 23, no. 3 (2016): 634, 643. As she puts it: "This [fated form of the weird] is, of course, a kind of weirdness permeating 1890s fiction, for instance, and one that is gaining renewed currency on the permeable borders of the contemporary realist novel" (Ibid., 643).

18. Henry David Thoreau, "Walking," in *Essays and Other Writings of Henry David Thoreau*, ed. Will H. Dircks (London: Walter Scott Limited, 1895), 13.

19. Ibid., 14–15.

20. See, for example, David R. Williams, *Wilderness Lost: The Religious Origins of the American Mind* (Selinsgrove: Susquehanna University Press, 1987), 173.

21. Jack London, *The Call of the Wild, White Fang, and Other Stories* (New York: Penguin, 1981), 125–26.

22. Gary Snyder, *The Practice of the Wild* (1990; Berkeley: Counterpoint, 2010), 16.

23. Ibid., 7.

24. Ibid., 20.

25. Marshall, "The Old Weird," 637.

26. Snyder, *The Practice of the Wild*, 17, 16.

27. Ibid., 9.

28. Fisher, *The Weird and the Eerie*, 13.

29. Snyder, *The Practice of the Wild*, 10–11.

30. Lovecraft, "Supernatural Horror in Literature," 107.

31. Halberstam, "Wildness, Loss, Death," 141.

32. Marshall, "The Old Weird," 637.

33. See Fisher, *The Weird and the Eerie*, 58; and Halberstam, "Wildness, Loss, Death," 145–47.

34. Jamaica Kincaid, "My Mother," in *The Weird: A Compendium of Strange and Dark Stories*, ed. Ann and Jeff VanderMeer (New York: Tor, 2012), 519.

35. Ibid.

36. Joyce Carol Oates, "Family," in *The Weird: A Compendium of Strange and Dark Stories*, ed. Ann and Jeff VanderMeer (New York: Tor, 2012), 761.
37. Ibid.
38. Michael Calia, "Writer Nic Pizzolatto on Thomas Ligotti and the Weird Secrets of 'True Detective,'" *The Wall Street Journal*, accessed May 6, 2019, https://blogs.wsj.com/speakeasy/2014/02/02/writer-nic-pizzolatto-on-thomas-ligotti-and-the-weird-secrets-of-true-detective/.
39. Thomas Ligotti, *Songs of a Dead Dreamer and Grimscribe* (London: Penguin Books, 2015), 3.
40. Ibid., 18.
41. Georges Bataille, "Base Materialism and Gnosticism," in *Visions of Excess: Selected Writings, 1927–1939*, ed. Allan Stoekl (Minneapolis: University of Minnesota Press, 1985), 45–52.
42. See, for instance, Paul Kingsnorth's and Dougald Hine's UK-based "Dark Mountain Project" and their 2009 manifesto, *Uncivilisation: The Dark Mountain Manifesto*. While this movement is British, *Uncivilisation* relies on Emerson's transcendentalism and Robinson Jeffers' modernist regionalism.
43. Halberstam, "Wildness, Loss, Death," 147.
44. See *OED*, "Outré," *Oxford English Dictionary*, accessed May 15, 2019.
45. Edgar Allan Poe, "The Murders in the Rue Morgue," in *Selected Tales* (London: Penguin, 1994), 134–45. I would like to express my thanks to Hanjo Berressem for reminding me of Dupin's word choice in this passage. In actual fact, the fictional detective goes on to describe "[t]he *wild* disorder of the room; the corpse thrust, with the head downward, up the chimney; the frightful mutilation of the body of the old lady... It is by these deviations from the plane of the ordinary, that reason feels its way, if at all in its search for the true" (Ibid., 135; emphasis added). Arguably, it is in this "*if at all*" that the genre of the weird may be located, as in something like a germinal state.
46. Thomas Ligotti, *The Conspiracy Against the Human Race: A Contrivance of Horror* (2010; London: Penguin Books, 2018), 171.
47. Ligotti, *Songs of a Dead Dreamer and Grimscribe*, 18.
48. Lovecraft, *At the Mountains of Madness: The Definitive Edition* (New York: The Modern Library, 2005), 102.

BIBLIOGRAPHY

Bataille, Georges. "Base Materialism and Gnosticism." In *Visions of Excess: Selected Writings, 1927–1939*, edited by Allan Stoekl, 45–52. Minneapolis: University of Minnesota Press, 1985.

Billy, Ted. "Descendentalism and the Dark Romantics: Poe, Hawthorne, Melville, and the Subversion of American Transcendentalism." In *A Companion to American Gothic*, edited by Charles L. Crow, 151–163. Oxford, UK: Wiley, 2013.

Calia, Michael. "Writer Nic Pizzolatto on Thomas Ligotti and the Weird Secrets of 'True Detective.'" *The Wall Street Journal*. Accessed May 6, 2019. https://blogs.wsj.com/speakeasy/2014/02/02/writer-nic-pizzolatto-on-thomas-ligotti-and-the-weird-secrets-of-true-detective/.

Cronon, William. "The Trouble with Wilderness: Or, Getting Back to the Wrong Nature." *Environmental History* 1, no. 1 (1996): 7–28.

Fisher, Mark. *The Weird and the Eerie*. London: Repeater Books, 2017.

Halberstam, Jack. "Wildness, Loss, Death." *Social Text* 32, no. 4 (2014): 137–148.

Kincaid, Jamaica. "My Mother." In *The Weird: A Compendium of Strange and Dark Stories*, edited by Ann and Jeff VanderMeer, 518–520. New York: Tor, 2012.

Ligotti, Thomas. *The Conspiracy Against the Human Race: A Contrivance of Horror*. 2010. London: Penguin Books, 2018.

———. *Songs of a Dead Dreamer and Grimscribe*. London: Penguin Books, 2015.

London, Jack. *The Call of the Wild, White Fang, and Other Stories*. New York: Penguin, 1981.

Lovecraft, H. P. *At the Mountains of Madness: The Definitive Edition*. New York: The Modern Library, 2005.

———. "Supernatural Horror in Literature." In *At the Mountains of Madness: The Definitive Edition*, 103–173. New York: The Modern Library, 2005.

Marshall, Kate. "The Old Weird." *Modernism/modernity* 23, no. 3 (2016): 631–649.

Oates, Joyce Carol. "Family." In *The Weird: A Compendium of Strange and Dark Stories*, edited by Ann and Jeff VanderMeer, 756–764. New York: Tor, 2012.

OED. "Outré." *Oxford English Dictionary*. Accessed May 15, 2019.

Poe, Edgar Allan. "The Murders in the Rue Morgue." In *Selected Tales*, 118–153. London: Penguin, 1994.

Snyder, Gary. *The Practice of the Wild*. 1990. Berkeley: Counterpoint, 2010.

Spaulding, Todd. "The Emerge(d)nt Weird Tale: A Genre Study." *Studies in the Fantastic* 3 (2015/2016): 76–99.

Thoreau, Henry David. "Walking." In *Essays and Other Writings of Henry David Thoreau*, edited by Will H. Dircks, 1–32. London: Walter Scott Limited, 1895.

Todorov, Tzvetan. *The Fantastic: A Structural Approach to a Literary Genre*. Translated by Richard Howard. 1973. New York: Cornell Paperbacks, 1975.

Wark, McKenzie. *Molecular Red: Theory for the Anthropocene*. London: Verso, 2015.

Weinstock, Jeffrey A. "The New Weird." In *New Directions in Popular Fiction: Genre, Distribution, Reproduction*, edited by Ken Gelder, 177–199. London: Palgrave Macmillan, 2016.

Williams, David R. *Wilderness Lost: The Religious Origins of the American Mind*. Selinsgrove: Susquehanna University Press, 1987.

CHAPTER 5

Queering the Weird: Unnatural Participations and the Mucosal in H. P. Lovecraft and Occulture

Patricia MacCormack

This chapter will explore the means by which queer and feminist readings of H. P. Lovecraft but against Lovecraftian purpose can offer new lines of flight toward creating terrains of the weird which open spaces for anti-anthropocentric and ahuman becomings. From Gods and Monsters to becoming imperceptible, indulging in Lovecraft's most intimate fears, including those of desire, gender, and flesh, create positive chaos and redeem both the weird and Lovecraftian fantasy—too long a world belonging to the hysteric, racist, sexist, and masculine—in ways that prove the integrated human has always been its own violent fantasy in need of debunking. Through the works of Gilles Deleuze, Félix Guattari, and what they call "unnatural participations,"[1] and Luce Irigaray on "mucous" and angels, and via some examples o fstrange hybrid couplings in Lovecraft, weird art, and feminist occulture, this chapter celebrates what Lovecraft most feared as a future for the ahuman.

P. MacCormack (✉)
Anglia Ruskin University, Cambridge, UK

© The Author(s) 2019
J. Greve and F. Zappe (eds.), *Spaces and Fictions of the Weird and the Fantastic*, Geocriticism and Spatial Literary Studies,
https://doi.org/10.1007/978-3-030-28116-8_5

Beyond Human

In Lovecraft's work many critics acknowledge a division between the mythos works and the horror works. I have elsewhere[2] demarcated instead six "gates" which reflect the six forms of unnatural alliances i n which Lovecraft's tediously white desireless male protagonists find themselves. The first relations concern animals and demons (for example in "The Beast in the Cave" [1918], "The Rats in the Walls" [1923], Brown Jenkins in "Dreams in the Witch House's" [1932] and the demons such as Azathoth and Nyarlathotep of his poetical works), the second the shift from fated heredity to abstract (not always consensual) alliance (Charles Dexter Ward and the fish folk of Innsmouth), the third inhuman becomings ("The Dunwich Horror" [1928], "The Silver Key" [1926]), the fourth aperceptive manifestation ("Beyond the Wall of Sleep" [1919], "The Call of Cthulhu" [1926]), the fifth from the horror of the unspeakable to the reality of the unsayable ("The Music of Eric Zann" [1921], *The Shadow Out of Time* [1936]) and the final gate takes all manifestations of matter, time, space, earth, and the cosmos beyond folds of perception to absolute molecular oneness (most explicitly in "Through the Gates of the Silver Key" [1934]). I cite these gates to show that Lovecraft's weirdness consists neither in the nihilism claimed by the male hysteric reader, nor in the nebulous horror encountered by that subject with something to lose, but rather to set up the way in which, for a queer reading, Lovecraft's work could be considered an escape route, in a very geologically ordered fashion, from the strictures of normative relations. Of course, though his protagonists are empathically white men, Lovecraft is a sexless writer and so any relations, queer or otherwise, are absent in his work. But not in my reading. Rather, these relations both shatter the fantasy of a hermeneutic and singular expression of subjectivity (to the point where some of his protagonists, such as Pickman, don't even need an other to create a multiple, fractal relation within themselves and between themselves, a schizo-becoming) and form a new geology of desire through encounters with alterity. The encounters must be queer because the alterity in Lovecraft is unspeakable and inconvertible to language or even form, where description shifts from hybridity to utter failure, so there is no oppositional or observable other. And there is an earthiness (or cosmic geology perhaps) to Lovecraft where, even while the entities may be irreducibly other both in perceptibility and potential to evaluate the nature of the protagonist's relation with them, the experiences of encounters are always deeply embedded in a fleshy materiality. Even time, through both the generations of Ancient Ones and Elder Gods, and the dubious heredity of Ward or the Innsmouth folk, seems to inhabit the cosmos with a laden materiality, extricating it from any transcendental metaphysical contemplation or esoteric absolute. Material viscerality is usually trite and necessary for horror but when time itself becomes visceral then the horror has reached a more profound level.

One persistent criticism of the weird is that it is often a reactive genre created for and by white men in the throes of existential angst (see for example Packer and Stoneman's history of pessimism from Socrates to Ligotti[3]). While not every weird author is heteronormative, one could suggest that they belong to a world that is homosocial or hom(me)osexual in that the weird has traditionally avoided including meaningful others of alternate genders, races and so forth. While for every claim that risks this ignorance there will be emergent examples that disprove it, because I am working with Lovecraft I will stand by the claim tactically, primarily not because of his work itself, but mainly due to the commentators who tend to either critique Lovecraft's absenting of women and explicit racism, or attend to his summary of the absolute insignificance of the human, always defaulted to the white man. If nothing else these men are the victims of the tales, and there is rarely an emerging victorious hero. This is precisely why I retain affection for Lovecraft's work—he celebrates rather than commiserates the end of the dominant human as we know it. In Lovecraft the message seems to adapt, evolve, become, it speaks with the voice of the outsider (most empathically and sympathetically in "The Outsider"). The tales also revel in the utter inconsequentiality of humans.

For a geological reading of the weird, which is a queer bedfellow with the ecological philosophies of environmentalism and the various human extinction advocacy movements, these tales can be seen as a positive, affirmative, and optimistic exploration of what the world would look like if white male capitalism wasn't destroying it. Like many science fiction and fantasy tales, as well as some writings in feminism and animal rights discourse, Lovecraft's Ancient Ones and Elder Gods stories ask the simple question: "Imagine a world where another species treated you how you treat women/animals/the Earth/those with less power?" But these are far from allegorical scoldings. The others, be they Ancient Ones or even human-faced rats in conjugal relations with witches, are not really that concerned with whether they destroy us or not. In a perverse opposition to equality politics, which asks the other to perform majoritarianism in order to count as equal, Lovecraft's protagonists have to adapt to unpredictable, interspecies, interkingdom, eternal-yet-metamorphic creatures and species, and their ways. His work is all about never knowing with what you will enter into a becoming, and the choice between revolutionary evolution or humanity as atrophy. In this sense Lovecraft offers the escape routes Gilles Deleuze and Félix Guattari claim are the essence of a minor literature: "to stake out a path of escape...to find a world of pure intensities where all forms come undone, as do all the significations, signifiers, and signifieds, to the benefit of an unformed matter of deterritorialized flux, of nonsignifying signs."[4] One can almost open any Lovecraft omnibus at any page to find not only this form of writing as content but also this form of expression as the inability to contain: "amidst [both through and around] backgrounds of other planets and systems and galaxies and cosmic continua; spores of eternal life drifting from world to world, universe to universe, yet all equally himself... His self had been annihilated and yet he—if indeed there could, in view of that utter nullity of individual existence, be such a thing as he – was equally aware of being in some inconceivable way a legion of selves."[5]

That minor literature is affiliated with becoming-woman and becoming-animal is seen with a strange redemptive resonance in Lovecraft's "Poetry and the Gods" where we see a rare female protagonist, Marcia, finding herself out of synch and in disdain of the human world, human language, the prosaic essence of humanity itself and even repulsed by human understandings of time. Gods and the cosmos, accessed through dreams and poetry become her muse, even while she sits aside a poet who thinks (mistakenly) she is his muse, but does not understand that while he remains human he will never attain the gods or cosmos. Marcia willingly seeks what many of Lovecraft's protagonists either resist, or are configured as arcane and evil for pursuing—the queer relation between unthinkable words and worlds, the delivery offered by dreams that reality and the impulse to logocentrism in its privilege of human exceptionalism denies. While Marcia is not described as a witch, she invokes the gods through poetry and her denial of anthropocentric time and space, much like the witch Keziah Mason, so one is a witch and the other described in shimmering abstraction, both belong away from the world of men, prose, society, and ordinary relations. Becomings and delivery from this world are seen as coveted, jubilant, not the horror with which so much of the events in Lovecraft's stories are met.

UNNATURAL PARTICIPATIONS

Lovecraft's legacy of both misogyny and racism is attested to in a post-Barthes/Foucault reading crucial to reclaiming the horror of the mucosal alien as the wondrous liberator of desire. In a reactive yet fascinating lament verging on a call to censorship of Lovecraft, and citing some personal letters that contain Lovecraft's horror at racial alterity and especially hybridity Timothy Letteney states:

> Now, compare this [racist] quote to an excerpt from Lovecraft's tale "At the Mountains of Madness" in which arctic explorers uncover an ancient alien race: '...we looked up from those headless, slime-coated shapes to the loathsome palimpsest sculptures and the diabolical dot groups of fresh slime on the wall beside them – looked and understood what must have triumphed and survived down there in the Cyclopean water city...' (495–496). The letter and the excerpt both describe alien beings originating from the sea, slimecoated surfaces, and Cyclopean masonry. The aliens he describes in the letter are in fact American citizens, as opposed to the actual headless slimy aliens he describes in the story. The interrelatedness of Lovecraft's racism and his fiction is impossible to ignore.[6]

I have no interest in redeeming Lovecraft, but Letteney's own horror at Lovecraft's disgust at alterity in general and racial, sexual and fascinatingly the visceral corporeality of the abject oozing alien he encountered upon moving to New York in particular emphasizes that there are options in the works. No feminist or anti-racist would tolerate the author but the rise in the popularity of Cthulhu and its pantheon is precisely because the kind of squeamish, spoiled, entitled mommy's boy that Letteney emphasizes Lovecraft was is precisely the reader who sees these as horror stories. What if we who identify with the horror, make lovers of the monsters, seek to join them in chaos rituals and becomings of mucosal libidinality and a jubilant indulgence in the very alien-ism that Lovecraft despises and fears and that we minoritarians have always been ascribed as? Like queer, we are taking Lovecraft's creatures away from his insulting intent and reclaiming them as our own, including the absolute disinterest the monstrous other shows to the kind of entitled white man that still rules society and tedious readings of weird fiction denied its inherent queerness in favor of existential egoism.

The witch logic of Marcia and Keziah, these two, rare women in Love-craft, leads to a contemporary extension of Lovecraft's world that has sim-ilarly been caught between a conservative understanding and a queer dissi-pation. Modern chaos magick, inspired originally by the art and literature of Austin Osman Spare and access to the unconscious, orients itself around a belief in the unbelievable or, more correctly, the untrue. Unlike religion, chaos magick knows that belief is a creative force reliant on no essential, unmalleable exteriority such as God, capitalism or any other contemporary religion. Chaos magick has appropriated the apocryphal world of Love-craft's Ancient Ones and Elder Gods, via the various versions of the fabled *Necronomicon*, the ancient grimoire that is also acknowledged as a 1970s' fake. The connection between Spare and Lovecraft is tenuous, but both are catalyzed by two trajectories. The first is a presumably Nietzschean inspired idea that "man" is capable of more than "man" thinks he is capable of, not through the simple and infantile reading of the *Übermensch* as a tyrannically powerful man (that is, "overman"), but as an organism that has surmounted the limits of what is meant by and valued as human (ironically those traits the misreadings seem to laud—power as oppressive, force as aggressive, the attainment of perfection as a final form). Power in chaos magick is what Deleuze and Guattari call force throughout their writings: not, forceful imposition of power but rather unthinkable expressivity that changes the very nature of the organism by refusing humanity as a form of commit-ment to repetition or aspiration to a very rigid template. The conservative reading of chaos magick uses this *force* to attain and accrue—usually the traditional occult suspects of wealth or sex (as advocated by, for example, Peter Carroll[7])—while the queer aspect encourages dissipation for its own sake, an experiment in self as many selves (inspired mainly in the work of Phil Hine[8] who adamantly configures the Old Ones as beyond this world "undimensional and unseen" but with whom we should nonetheless enter into a becoming[9]).

The burgeoning fashionability of cephalopod sex both belongs and does not belong to queering the weird in occulture. While tedious manga porn designed for the male gaze repeats heteronormative and phallic patterns of sex and violence, there is a nonlinear geoplane of desire we can see from Hokusai's *Dreams of the Fisherman's Wife* (1815) where the octo-pus is duplicated to create a doubling between two beaks and two sets of lips on a woman, one oral and one labial, reflecting the work of Luce Irigaray on female sexuality and female discourse in *This Sex Which Is Not One*.[10] In the Hokusai print the unnatural nuptials defy any template of heteronormativity based on isomorphic binarizations of active/passive or penetrative/penetrating because they are interkingdom—that is the three participants belong to three different kingdoms yet they all libidinally inter-act, a natural unnatural participation. In this sense without male or phallic sexuality these binaries become redundant. Addressing this as more than just a titillating image, the metaphor of two lips, two mouths/beaks, and two sets of each shows the mutuality in discourse as an act of compressive reciprocity, a moment of mucosal discourse and mucosal love. The mucos-ity of the cephalopod, long considered belonging to the abject world of slime, female sexual fluid, and general alien otherness, is the moment of the being between, of sticky desire, of relations created by oozing connec-tivity rather than the vast chasm that separates oppositional discourse and oppositional desire based on speaker and listener, pedagogue and student, penetrator and penetrated or, in the psychoanalytic tradition, the phallic and the lacking (seen so emphatically in Picasso's dreadful "homage" to Hokusai in *Woman and Octopus*). These are all heterohumanist hangovers. Weird queer goes beyond these structures both discursively and desiringly, so where Lovecraft imagines we will flee from the Great Old Ones, our queer readings draw us to them for with them we can live in the mucous. The Great Old Ones do not find woman lacking, and are happy to gaze into the abyss that the nihilist, pessimist, or speculative realist sees in the other through their demand that the other define or announce themselves,

> ...it is only in an act that mucous perceives and loves itself without thesis, without position outside itself. The potency achieves "its" act which is never set in a finished piece of work. But which is always *half open*...The anxiety of the chasm, of the abyss, equally, on the part of the man who neither welcomes nor finds a rhythm for the act of love. This failure to embrace the mucous leads to the squandering of its abundance, the exploitation of its availability, its joyfulness, its flesh, or to the abandonment and repetition of its gesture or gestures of love, which become broken and jerky, instead of progressive and inscribed in duration.[11]

The becomings in Lovecraft as mingling mucosal dissipations defy the bro-ken and jerky sporadic qualities of anthropocentric time which is often the dismemberment of duration by rationalization, the same logic which sees the abyss as absence rather than voluminosity and the demarcation of self and other as the basic tenet of egoistic life. Other examples of cephalopodic becomings that both inspired Lovecraft before his writing and were inspired by the Ancient Ones include Fernard Khnopff's 1888 *Istar* where the goddess emerges (or submerges) in ecstasy with and as tentacles and connects with the vital role of Istar as Inanna in all versions of *The Necronomicon*. Andrej Zulawski's 1981 film *Possession* shows similar durations of becoming between a woman and her first mucosal, then vegetal/florescent, then tentacled lover, foiling both relations of desire and demarcations between lover and beloved as they create a univocal but poly-intensities emergence of love. In chaos magick the seminal images by Brian Ward to illustrate *Liber Null* and *Psychonaut* similarly show vulvic symmetry co-emergent with cephalopod gods directly inspired from Lovecraft but with a strangely enigmatic androgyny that recalls the androgyny of Baphomet that they also reference, and who as a feminine goat God appears frequently in the art of Spare.

The above-referenced artworks and the works of Biblical Apocrypha and Theosophy below show that we women, we minorities, we queers have been in a mucosal union with Cthulhu and his aeon well before Lovecraft and happily so.

Encounters of Ecstasy

A key element in mucosal interkingdom becomings implicit in the denial of time includes the denial of the cessation of act of love through any capitalist sense of satisfaction, which limits the act of love to a finite and consumable narrative. For this reason, I would prefer the term "ecstasy" in describing the affectivity of these relations, in the works of art and cinema, in chaos rituals, and in Lovecraft. Ecstasy is qualitatively similar to teratology in that it is independent of binaries of good/bad or pleasure/pain (just as *teras*'s meaning of fascination includes both horror and wonder without evaluation). Ecstasy in many forms of art collapses not simply binaries but the capacity for expression and language themselves to enclose experience. The silent speechless horror and the confounding sounds, images, and glyphs in Lovecraft's tales are never translated, the apocryphal books do not exist. There are events and languages but they are not apprehensible through anthropocentric modes of perception.

Mucosal modes summon gods. Connecting her work with a vitalistic reading of Nietzsche, Irigaray states: "In other words, because the mucous has special touch and properties, it would stand in the way of the transcendence of a God of immutable, stable truth. On the contrary, the mucous would summon the god to return or to come in a new incarnation, a new parousia."[12] The clearest example of such a summoning is seen in Bernini's two ecstasy sculptures, *The Ecstasy of Saint Teresa* (1652) and *The Ecstasy of Blessed Ludovica Albertoni* (1674). The former is particularly apt as it resonates with the weird fiction of Saint Teresa of Avila's own diaries (circa 1567),[13] which read like a Lovecraft story of becoming but with joy rather than horror. The one liberty I take with Teresa as sculpture and diarist is that I would reconfigure her experience with the angel who visited her not as a messenger from God (who cannot live in mucous) but perhaps from "elsewhere"—the unnatural participation of Biblical apocrypha incarnated as angel sex. Teresa seems to indulge in strange delight in describing the joys and pleasures the Devil brings while simultaneously maligning them as ephemeral.[14] For Teresa her angel-induced ecstasy is pure and eternal. In one of her most mystical and corporeal ideas Irigaray states: "The consequences of such non-fulfilment of the sexual act remain... to take only the most beautiful example... let us consider the angels. These messengers are never immobile nor do they ever dwell in one single place. As mediators of what has not yet taken place... these angels therefore open up the closed nature of the world, identity, action, history."[15] We cannot forget—especially because of the crucial role of the Watchers in their incarnation in the 1979 Simon version of *The Necronomicon* and their earliest incarnation in a founding text of chaos magick, *The Book of Enoch*, written and first published around 300 BCE (arguably both apocryphal texts, the latter ancient, the former thoroughly modern)—that angels fall. And when they fall, they fall not for humans, but for women, as seen in Genesis 6:4 which describes the birth of the Nephilim, and Book 6 of Enoch. Strangely the devil is not considered one of these interstitial beings, but the devil's incarnation as Baphomet in contemporary occulture suggests they all share certain qualities—they transform, they occupy a half-space (halfway between heaven and earth or earth and hell), they are androgynous, they are most interested in women rather than humans, and they stand against the authoritarianism of a single Godhead, which can be described as God but which can also be understood as the phallogocentric limits of majoritarian white hetero male anthropocentrism—man who made God in his own image.

What connects the biblical Nephilim with Lovecraft is the persistence of their existence beyond time, yet also in the present. They are of, from, with, and as an ancient race, the idea we belong in a covenant with an unseen and unseeable genealogy. In Lovecraft it invokes through monsters such as Cthulhu, and in *The Necronomicon* ancient Sumerian gods such as Pazuzu and Erishkegal. Alliance through becomings with demons/angels creates both communal or shared folds between (threshold) as well as new folds within the singular self (an alternation of Deleuze's, after Leibniz's *habitus*). This necessitates alteration in perceptions through alterations in being, the threshold of which perhaps we could describe, as Lovecraft does in "The Unnamable" as a "hybrid nightmare"[16] (the precise insult he also hurls at non-WASP Americans).

Angels, be they Watchers or the Devil (in his guises as Lucifer, Satan or simply the antihero), belong to the worlds of mucosal monsters, and the relations of desire they afford are transporters, be they through knowledge as in Enoch or Genesis or desire in hybrid cosmic becomings. Deleuze and Guattari tell us: "the devil is a transporter."[17] In ecstasy the encounter transports the self, rather than being a means by which the self is reaffirmed, which is how the sexual narrative of capitalism—heteronormativity—structures desire via separation of object and subject and suppression of the mucous. Both self and other (if they are even describable as such in their coalescent co-emergence) are transporter and transported, entering with openness and emerging with a union that is outside of time and within a space beyond context or definition.

HEAVENLY BRIDEGROOMS

Woman, that nebulous entity both ignored and possibly despised by Lovecraft, has long been aligned with those who transport. Recently authors such as Robert Lima[18] and Per Faxneld[19] have traced the allegiance between women, evil, pre-Christian witches, apocryphal rebel angels and Satan, but this reaches back to Mary Wollstonecraft (1792)[20] and the early Victorian tendency to positively align the feminine with Milton's fallen hero. Far from being in opposition to evil, these unnatural participations with otherworldly beings resonate with the Heavenly Bridegrooms described during the Middle Ages to apprehend the unique relationship between Saints and the Holy Spirit, representative of the material nature of the mystic and their (usually *her*) experiences. While such ecstasies were sometimes mistaken as Loudon-esque possessions, both aspects of the coupling were seen as viable and, most importantly, material, predating the Cartesian division between mind and flesh and recalling the fleshly ascendance of Mary and Enoch without conversion to soul or spirit. This brief introduction to a form of unnatural participation is not simply to connect the Watchers of the apocrypha with those of the *Necronomicon* but to introduce the vastly under-addressed author Ida Craddock and her *fin-de-siècle* work on both marital relations and her own marital relations with her heavenly bridegroom, an astral spirit with which she had sexual relations, the subject of which is her magnificent treatise of the same name (1894) in which she traces the impossibility of dividing angels or visiting creatures into good and evil, angel or demon, monster or holy spirit. Her reading of medieval investigations o fthe apocrypha is eerily reminiscent of Lovecraft. She begins by emphasizing that the entire history of woman as temptation is a patriarchal myth as in apocrypha it is first the Grigori angels themselves blamed for the depravity of their union with human women but ultimately "it was not the wickedness of the angels who wedded earthly women, but the evil imaginations of the human heart that brought about the punishment of the deluge."[21] Craddock's reading puts the blame squarely on men for wickedness, aligning women and the fallen Watchers because "man was pre-eminently the sex which violated the laws of right living."[22]

The interstitial or in-between nature of these fallen angels (perhaps explaining the difference between the Watchers and Satan's pandemonium) comes from Craddock's meticulous research on early medieval scholars, for example where she summarizes Lactantius stating: "Thus from angels the Devil makes them to become his satellites and attendants. But they who were born from these [the Nephilim], because they were neither angels nor men, but bearing a kind of mixed (middle) nature, were not admitted into Hell as their fathers were not into Heaven. Thus there came to be two kind of *demnos,* one of Heaven, the other of Earth."[23] This generational division of gods and monsters resonates directly with Lovecraft's two generations of Elder Gods and Ancient Ones, but in Craddock's work we see that women belong to this order of *demnos,* or "Moonly" beings, seen in her connection between the Talmud's Lilit with the Greek Mood Goddess Ilythlia, the Roman Alilat as the Star of Venus, while man alone is terrestrial in his concerns and relations. Craddock connects these interstitial angels with the heaven and earthborn archaic Dionysus and Hellenic Bacchus and also in the angelic capacity as messenger with Mercury, while the progeny of unnatural participations produce in early Islam the Shiqq, "another form of the Giant progeny of Borderland Unions."[24] Craddock's entire book traces a sensitive and scholarly trajectory between the persistent alliance of the in-between nature of women, including their between human and monster and their between spirit and (carnal) flesh and their denigration as evil, while all along reminding us this book is not a historical testament but a manual on how to invoke heavenly bridegrooms, firmly set in her practical magic theosophical tradition.

What is most tragic is Craddock was confined to asylums by her mother during her life and ultimately took her own life as a result of being charged and sentenced to imprisonment for circulation of obscene literature due to two of her marriage works, "The Wedding Night" and "Right Marital Living" (both originally published in 1900). Both are explicit but not titillating guides advocating nonreproductive sex (primarily oral for both parties), a form of what would now be called tantric sex (retainment of semen) and the necessary spiritual nature of carnality. She states: "But the husband and wife who have known the bliss of controlled orgasm [of the man]…enter to find themselves, as it were, at the very Heart of those forces which first sent the nebulous mass of our solar Universe whirling into space. They are in Chaos, but a Chaos which is being evolved into a Cosmos. They struggle in the foaming rapids of sexual creative passion…."[25] Compare this to the very concept of "cosmic horror," "The Crawling Chaos," which Lovecraft wrote with Winifred Jackson, Cthulhu's "nebulously recombining" eventually achieves "eldritch contradictions of all matter, force and cosmic order"[26] and the "teeming, seething, swelling, foaming of Nameless Horror"[27] Deleuze and Guattari quote Lovecraft as dreading but they advocate as essential for a phenomenon of bordering that constitutes becomings. One cannot help but wonder if Craddock, writing in New York, where she committed suicide in 1902, was not part of the literary legacy that so disgusted Lovecraft, and yet both admit a double pantheon, albeit one who apprehends it with horror and the other with desire. It may seem disrespectful to address Craddock's serious theosophical working as a form of queer desire in weird literature, but her resonances with Lovecraft's intensities and geoplanes of Gods and Monsters, spirit and flesh makes her invaluable in elucidating that the world of the in-between and unnatural participations can be as liberating for some as it is horrifying for others. She is an author who will hopefully reemerge in popularity as a great scholar both of the occult and of desire.

AGAINST NATURE, FOR NATURE

"When their union with the supposed Devil was based on the faithful tender love of one woman for one man and its reciprocity, in accordance with higher moral standards, then was the union objective and natural."[28] So states Craddock on all participations, showing the amoral, Nietzschean nature of nature itself, which, as Deleuze and Guattari remind us: "These combinations are neither genetic nor structural; they are interkingdoms, unnatural participations. That is the only way Nature operates—against itself."[29]

There is always something deeply ecological and simultaneously deeply alien about nature and its infinite rhizomatic connection with the cosmos as dramatized in and by weird fiction. This feeling of alienness or "wrongness" which the usual suspects of speculative realism and weird criticism invoke is one of utterly normal alienation for minoritarians, for the feminine and for all others who do not wish to have to confess their alterity through anthropocentric language in order to be realized to the dominant, no matter with how much tentative speculation. The inapprehensible other owes the dominant nothing—no language, no meaning, certainly no ontological manifestation. The other finds its own among the unlike-itself in nature, the infinite chaos that nature offers well in excess of any taxonomy the social contract between men among men can glean. When art, in all of its different incarnations, is introduced into the mix we have even more opportunities to see the unnatural participations of nature against itself. The weird can retain its weirdness for both those who encounter it with horror or with desire, but queering the weird welcomes the alien. In this sense, i t resonates with the reimaginings of the occult and particularly the figure of the witch in contemporary occulture among feminists. Through unnatural readings of Lovecraft and remembering lesser celebrated but clearly in their time influential writers such as Craddock, queering the weird delivers it from its gender bifurcations, its unpalatable racist and sexist implications, and connect it with its own unnatural participations that open the world to strange combinations and infinite relations without hierarchy, taxonomy, classification toward an ecological plateau of queer cosmic desire.

Notes

1. Gilles Deleuze and Félix Guattari, *A Thousand Plateaus: Capitalism and Schizophrenia*, vol. 2, trans. Brian Massumi (London: Athlone, 1987), 240.
2. See Patricia MacCormack, "Lovecraft Through the Deleuzio-Guattarian Gates," *Postmodern Culture* 20:2 (2010).
3. Joseph Packer and Ethan Stoneman, *A Feeling of Wrongness: Pessimistic Rhetoric on the Fringes of Popular Culture* (Pittsburgh: Penn State University Press, 2018).
4. Gilles Deleuze and Félix Guattari, *Kafka: Toward a Minor Literature*, trans. Dana Polan (Minneapolis, MN: University of Minnesota Press, 1986), 13.
5. "Through the Gates of the Silver Key," *H.P. Lovecraft Omnibus 1* (London: Voyager, 1999), 526–27.

6. Timothy Letteney, "I Just Called to Say Cthulhu: Xenophobia and Antiquarianism in H.P. Lovecraft's Mythos," 4–5, accessed May 15, 2019, https://www.academia.edu/4144322/I_Just_Called_to_Say_Cthulhu_Xenophobia_and_Antiquarianism_in_H.P._Lovecraft_s_Mythos?fbclid=IwAR0audNgMGqyt9pKYzys9xXNksc4y_dOXJM4zm9OXj0h_ILnQoZeB49c7QU.

7. Peter Carroll, *Liber Null and Psychonaut* (York Beach, ME: Weiser, 1987).

8. Phil Hine, *Prime Chaos* (Tempe, AZ: Falcon Press, 1993) and *The Pseudonomicon* (Tempe, AZ: Falcon Press, 2004).

9. Hine, *Pseudonomicon*, 17.

10. Luce Irigaray, *This Sex Which Is Not One*, trans. Catherine Porter (Ithaca: Cornell University Press, 1985).

11. Luce Irigaray, *An Ethics of Sexual Difference*, trans. Carolyn Burke and Gillian C. Gill (Ithaca: Cornell University Press, 1993), 111.

12. Ibid., 110.

13. Teresa of Avila, *The Life of Saint Teresa of Avila by Herself*, trans J.M. Cohen (London: Penguin, 1957).

14. Ibid., 178.

15. Luce Irigaray, *The Irigaray Reader*, trans. Margaret Whitford (London: Blackwell, 1992), 173.

16. H.P. Lovecraft, "The Unnamable," in *H.P. Lovecraft Omnibus 2* (London: Grafton, 1985), 232.

17. Deleuze and Guattari, *A Thousand Plateaus*, 253.

18. Robert Lima, "Gendering Evil: Pandora, Lilith, Satan," in *The Iconology of Gender I*, eds. Attila Kiss and Georg Szonyi (Szeged: University of Szeged Press, 2008).

19. Per Faxneld, *Satanic Feminism: Lucifer as Liberator of Woman in Nineteenth Century Culture* (Oxford: Oxford University Press, 2017).

20. Mary Wollstonecraft, *Vindication of the Rights of Women* (London: Penguin, 1975).

21. Ida Craddock, "Heavenly Bridegrooms," in *Ida Craddock Collection* (McAllister Editions, 2017), 8.

22. Ibid., 12.

23. Ibid., 15.

24. Ibid., 46.

25. Ibid., "Right Marital Living," in *Ida Craddock Collection* (McAllister Editions, 2017), 179.

26. H.P. Lovecraft, "The Call of Cthulhu," in *H.P. Lovecraft Omnibus 3* (London: Grafton, 1989), 97, 99.

27. Deleuze and Guattari, *A Thousand Plateaus*, 245.

28. Craddock, "Heavenly Bridegrooms," 65.

29. Deleuze and Guattari, *A Thousand Plateaus*, 242.

Bibliography

Carroll, Peter. *Liber Null and Psychonaut*. York Beach, ME: Weiser, 1993.

Charles, R.H. *The Book of Enoch the Prophet*. York Beach, ME: Weiser, 2003.

Craddock, Ida. "Heavenly Bridegrooms." In *Ida Craddock Collection*, 1–95. McAllister Editions, 2017.

———. "Right Marital Living." In *Ida Craddock Collection*, 161–77. McAllister Editions, 2017.

Deleuze, Gilles, and Félix Guattari. *Kafka: Toward a Minor Literature*. Translated by Dana Polan. Minneapolis: University of Minnesota Press, 1986.

———. *A Thousand Plateaus: Capitalism and Schizophrenia*, vol. 2. Translated by Brian Massumi. London: Athlone, 1987.

Faxneld, Per. *Satanic Feminism: Lucifer as Liberator of Woman in Nineteenth Century Culture*. Oxford: Oxford University Press, 2017.

Hine, Phil. *Prime Chaos*. Tempe, AZ: Falcon Press, 1993.

———. *The Pseudonomicon*. Tempe, AZ: Falcon Press, 2004.

Irigaray, Luce. *An Ethics of Sexual Difference*. Translated by Carolyn Burke and Gillian C. Gill. Ithaca: Cornell University Press, 1993.

———. *The Irigaray Reader*. Translated by Margaret Whitford. London: Blackwell, 1992.

Letteney, Timothy. "I Just Called to Say Cthulhu: Xenophobia and Antiquarianism in H.P. Lovecraft's Mythos." Academia.edu. Accessed March 3, 2018. https://www.academia.edu/4144322/I_Just_Called_to_Say_Cthulhu_Xenophobia_and_Antiquarianism_in_H.P._Lovecraft_s_Mythos?fbclid=IwAR0audNgMGqyt9pKYzys9xXNksc4y_dOXJM4zm9OXj0h_ILnQoZeB49c7QU.

Lima, Robert. "Gendering Evil: Pandora, Lilith, Satan." In *The Iconology of Gender I*, edited by Attila Kiss and Georg Szonyi, 9–20. Szeged: University of Szeged Press, 2008.

Lovecraft, H.P. *Supernatural Horror in Literature*. New York: Dover, 1973.

———. "Dagon." *H.P. Lovecraft Omnibus 2*, 11–17. London: Grafton, 1985.

———. "Poetry and the Gods." *H.P. Lovecraft Omnibus 2*, 382–90. London: Grafton, 1985.

———. "The Nameless City." *H.P. Lovecraft Omnibus 2*, 129–43. London: Grafton, 1985.

———. "The Unnamable." *H.P. Lovecraft Omnibus 2*, 225–35. London: Grafton, 1985.

———. "Beyond the Wall of Sleep." *H.P. Lovecraft Omnibus 2*, 36–48. London: Grafton, 1989.

———. "The Call of Cthulhu." *H.P. Lovecraft Omnibus 3*, 61–98. London: Grafton, 1989.

———. "The Colour Out of Space." *H.P. Lovecraft Omnibus 3*, 236–71. London: Grafton, 1989.

———. "The Dunwich Horror." *H.P. Lovecraft Omnibus 3*, 99–153. London: Grafton, 1989.

———. "The Music of Eric Zann." *H.P. Lovecraft Omnibus 3*, 335–45. London: Grafton, 1989.

———. "The Shadow Out of Time." *H.P. Lovecraft Omnibus 3*, 464–544. London: Grafton, 1989.

———. "The Shadow Over Innsmouth." *H.P. Lovecraft Omnibus 3*, 382–463. London: Grafton, 1989.

———. "The Case of Charles Dexter Ward." *H.P. Lovecraft Omnibus 1*, 141–302. London: Voyager, 1999.

———. "The Dream-Quest of Unknown Kadath." *H.P. Lovecraft Omnibus 1*, 361–486. London: Voyager, 1999.

———. "The Dreams in the Witch-House." *H.P. Lovecraft Omnibus 1*, 303–50. London: Voyager, 1999.

———. "Through the Gates of the Silver Key." *H.P. Lovecraft Omnibus 1*. London: Voyager, 1999.

———. "Azathoth." *The Ancient Track: The Complete Poetical Works of H.P. Lovecraft*, 73. San Francisco: Nightshade Books, 2001.

———. "Nyarlathotep." *The Ancient Track: The Complete Poetical Works of H.P. Lovecraft*, 72–73. San Francisco: Nightshade Books, 2001.

MacCormack, Patricia. "Lovecraft Through Deleuzio-Guattarian Gates." *Postmodern Culture* 20:2 (2010). http://www.pomoculture.org/2013/09/03/lovecraft-through-deleuzio-guattarian-gates/.

Simon. *The Necronomicon*. New York: Avon Books, 1977.

Teresa of Avila. *The Life of Saint Teresa of Avila by Herself*. Translated by J.M. Cohen. London: Penguin, 1957.

Wollstonecraft, Mary. *Vindication of the Rights of Women*. London: Penguin, 1975.

CHAPTER 6

Geological Insurrections: Politics of Planetary Weirding from China Miéville to N. K. Jemisin

Moritz Ingwersen

Roughly one year after the announcement of the Anthropocene by meteorologist Paul J. Crutzen, the science fiction writer and literary critic M. John Harrison jump-started a discussion about the emergence of a new literary genre—the New Weird.[1] Arguably, the referents of neither were new, yet both terms emerged more or less contemporaneously to provide an umbrella for envisioning the relationship between the human and the world as a function of simultaneously intimacy and estrangement. The trajectory from Jeff and Ann VanderMeer's collection *The New Weird* (2007), to Gerry Canavan and Andrew Hageman's special issue *Global Weirding* (2016), Donna Haraway's *Staying with the Trouble: Making Kin in the Chthulucene* (2016), and Anna Tsing's *Arts of Living on a Damaged Planet: Ghosts and Monsters of the Anthropocene* (2017) is emblematic of ongoing refractions of the current ecological crisis through the registers o f he

M. Ingwersen (✉)
University of Konstanz, Konstanz, Germany

© The Author(s) 2019
J. Greve and F. Zappe (eds.), *Spaces and Fictions of the Weird and the Fantastic*, Geocriticism and Spatial Literary Studies,
https://doi.org/10.1007/978-3-030-28116-8_6

weird and the uncanny. In light of the increasing frequency of seismic trembles, hurricanes, freak temperatures, and toxic spills, the world, it seems, is haunted by an ill-disposed return of what 300 years of industrial capitalism have gone to great lengths to repress: the agency and complexity of a planetary ecology that apart from the human will to dominate it harbors a multiplicity of inhuman[2] alliances that hover just below our sensory threshold, in the unexplored abyss just below the habitable surface, and in the infinitesimal cracks of a ground that the humanist tradition has relentlessly constructed as solid and reliable. Conspicuously, the proclamation of the Anthropocene both reasserts and contests this predicament. Appropriately or not, the human is validated as the presiding climatological force and at the same time displaced as it is embedded in the planetary sediment. Inversely, understanding the human as a geological subject also requires critical reevaluation of the ways in which agency, life-force, and geopolitical activity is distributed across a wide spectrum of human and inhuman actors. As Bruno Latour puts it, "Earth has become—has become again!—an active, local, limited, sensitive, fragile, quaking, and easily tickled envelope."[3] No longer the yielding object of the human gaze and prod, but increasingly unthinkable, "the world often 'bites back,' resists, or ignores our attempts to mold it into the world-for-us,"[4] Eugene Thacker writes. In his book *In the Dust of This Planet* (2011) he examines how supernatural horror fiction literalizes "the subtraction of the human from the world" as a prospect "that is at once impersonal and horrific."[5] In the age of the Anthropocene, modern fantasies of bounded individualism—formerly called upon to safeguard clear dividing lines between inside and outside, above and below, so-called nature and so-called culture—prove increasingly untenable. Instead, what we are compelled to come to terms with are porosities, hybrids, and insurrections from the ground up—uncanny specters that readers of weird fiction will be all too familiar with.

In this chapter, I want to suggest that it is specifically through its persistent imagining of geological confrontations, unsettlings, and hauntings that the tradition of weird fiction literalizes the geomorphic enmeshments of the Anthropocene. Illustrating the migration of geological metaphors from the Old to the New Weird, I will foreground the work of the award-winning fantasy writer N. K. Jemisin. Her programmatically titled *Broken Earth* trilogy (2015–2017) shall serve as an example of a type of geostory—to borrow a term from Bruno Latour[6]—in which the celebration of collaboration, kinship, and resurgence effectively subverts the xenophobic demonization of inhuman alterity which has become synonymous with the Old Weird. With outspoken contempt for the overt racism and patriarchal undertones in Lovecraft's work,[7] Jemisin has emerged as a lucid reinventor of many of the genre's recurring tropes.[8] My intention is not unequivocally to champion Jemisin as an author of the weird—a label she would almost certainly reject. Rather, I want to show how the planetary estrangements she depicts lend themselves to an ecocritical reading that may supplement and challenge recent receptions of both weird fiction and the Anthropocene. As I illustrate below, her work contests the apocalyptic specter of extinction or a "world-without-us"—identified by Thacker as a key element of weird climate anxieties—and instead unfolds an ethics of collaboration, entanglement, and what North American Indigenous scholars know as survivance.[9] If Jemisin's metabolization of the Anthropocene resonates with the dystopian vision of a world-without-us, it does so by foregrounding that "imperialism and ongoing (settler) colonialism have been ending worlds for as long as they have been in existence"[10] and that these histories are inextricably interwoven with the onset of resource capitalism.

THE GEOLOGICAL UNCANNY

It might not surprise that anxieties about a decentering of the human have shaped geological imaginaries since the nineteenth century. Evocatively, Stephen Jay Gould captures the unsettling implications of the discovery of geological time-scales by James Hutton and Charles Lyell:

> What could be more comforting, what more convenient for human domination, than the traditional concept of a young earth, ruled by human will within days of its origin. How threatening, by contrast, the notion of an almost incomprehensible immensity, with human habitation restricted to a millimicrosecond at the very end![11]

Arguably, this capitulation of the human in face of the immensity of time and space is precisely what lay at the heart of Lovecraft's vision of the weird. Through his invocation of "depths unimagined and unimaginable,"[12] "stupefying gulfs of time,"[13] "the prodigious geological upheavals of those primal days,"[14] or "the writhings of earth's crust,"[15] scenes of geological unsettlement have emerged as a preeminent trope of the genre.[16] The baseline of many of Lovecraft's tales is that Earth itself is an alien planet and that underneath the thin layer of its surface chaos and the ancient and unthinkable reign supreme. It is not difficult to perceive the proximity between this territorial anxiety and xenophobic, ableist, and racist ideologies of "ontology hygiene."[17] For Lovecraft, it seems, complexity, boundary transgressions, and cultural diversity are inherently frightening and suspicious—a conservatism that leads Jemisin to accuse him of allegorically "trying to gentrify every city he lived in."[18] When Michael Moorcock announces that "the appeal of the weird story is precisely that it is designed to disturb,"[19] contemporary readers are inclined to note that Lovecraft's visions do less to disturb than to reinforce the status quo and its normalized disdain for alterity and deviation. His politics seem like a direct extension of the imperialist program that has shaped the conceptual infrastructure of geology, which has since the nineteenth century contributed to the dehumanization and exploitation of racialized and nonnormative others via a universalized extension of extractivist ideologies. To truly disturb or disrupt the cherished delusion of human exceptionalism would mean to imagine geological agency and the transgression of ontological boundaries not as a threat but as an opportunity to produce more inclusive models of what it means to inhabit the climate crisis relationally and collectively, of reciprocal affection and "becoming-with"[20] rather than becoming extinct, of mutual "response-ability"[21] rather than unilateral superiority.

China Miéville's short story "The Dusty Hat" (2015) may serve as an example of the ways in which many writers of the New Weird have metabolized the predicaments of the Anthropocene through a progressive reenvisioning of inhuman agency and political participation. Miéville's settings frequently involve a defamiliarization of the environment that entails an upheaval of entrenched oppositions between surface and underground, life and nonlife. Not surprisingly for an author who self-identifies as a "revolutionary socialist,"[22] the ecological insurrections he stages occur from the bottom up: a return from oceanic depths[23] or the crevices of a dusty floor. Like Freud's uncanny, the unthinkable other comes from the interior; it has always been part of the system. In "The Dusty Hat," the commingling of human and inhuman worlds is cast in explicitly political registers. The story is framed in the form of a letter written by an unnamed narrator to explain his mysterious disappearance in the aftermath of a meeting of leftist activists. His account of the strange events that followed opens with the observation of cities afflicted with an inexplicable geological porosity: "There seemed to be sinkholes opening up everywhere. I was looking at pictures of cars angling up from where roads had subsided into nothing, giant holes in the cement of cities around the world."[24] It remains unclear, whether these perforations of reality are generally perceptible or a symptom of the heightened sensitivity to the world's materiality that accompanies the strange metamorphosis the hardly reliable narrator is about to describe. He recounts how, triggered by the encounter with an old man wearing a conspicuously dusty hat, his reality and customary sense of subjectivity have begun to shift. In a literal upheaval of his comfort zone, "[e]verything swayed"[25] as the mysterious man shows up on his doorstep after the conference; "[his] room pitched [and he] started to slide as if into a sea."[26]

Mirroring the proliferation of sinkholes, the material ground of his existence gives way and opens up to the dizzying revelation that every element of his inhuman environment is not only sentient but firing up for an inorganic revolution. In a grotesque material eversion, what the protagonist took to be a man reveals itself as a mere human shell to a socialist collective of dust. Ventriloquizing through the decomposing larynx of "the man the dust wore,"[27] it alerts the protagonist to "the politics of objects"[28] expressed in "arguments between [YouTube] images,"[29] "interactions of angles of furniture,"[30] and muttering floorboards. He learns that the fissure in his wall is "a loyalist crack,"[31] that "[a]rchitecture's always centrist,"[32] and that the dusty hat is a "viewing platform"[33] for "a scouting layer."[34]

It is at this point in the story that we learn about the nature of the narrator's enigmatic disappearance; he is recruited to a political fight on a new diversified front: "I inhaled the dust. In it rushed. My body must have thought I was dying. Probably I was writhing and twitching alongside the old skin. ... What I am trying to tell you—for which you may not thank me—is that the dust was and is my comrade."[35] Unified with the "revolutionary unliving,"[36] the protagonist becomes sensitive to an ancient political struggle of planetary proportions that dramatically reveals the forces at play in the world-without-us:

> Cycles of geological insurrection. Vaalbara, prelapsarian collectivity of stone and surface. Kenorland and Pangea, peace becoming war; the rage of the gap at the unbreached, totalities torqued apart over mere glimmering millions of years. A savaging of scale, Triassic wars of position as Gondwanaland and Laurasia rounded in ruthless continental pugilism, their own components in solidarity, plateaus heaving, shale slipping masses, subject-objects of history, scree in struggle against the bottomness of holes. A primitive communism of granularity, grassroots democracy before there was grass or roots or anything but hot dirt, until at last there were birds and an epoch of walls.[37]

The erosion of entrenched separations between human and inhuman subjectivities is reflected in the baroque interlinkage of customarily unrelated semantic registers. Miéville invokes "a communism of granularity,"[38] "[c]rumbling as syndicalism, the ca'canny of quartz,"[39] "[f]lint ultraleftism; dirt voluntarism; glass struggle."[40] Elsewhere, he evocatively, elevates the destabilization of geological foundations as the poetic mantra of the weird: "The ground below [our minds] is hole-y. There are cracks and chaos, meaningquakes. The metaphors we walk on are scree."[41]

In "The Dusty Hat," the Anthropocene is revisited as "a baroque new fascism of flesh"[42] that provokes the formation of a planetary environmental proletariat. The baseline of this literalized dialectical materialism in deep time is that "[n]one of us can stay. This epoch gets you coming or going."[43] The cyclical nature of this incessant transformation is recapitulated as much in the protagonist's volte in ontological and political affiliation as it is in the story's narratological structure. The plot opens at a leftist conference; it ends with an analogous but distinctly less conspicuous political meeting that invokes an urban dumping site but could be the incipient stage of what Latour envisions as a heterogeneous "parliament of things."[44] It takes place at

> the corner of canal, where sunken bikes and a rust-scaled supermarket trolley were visible through shallow waters below a half-melted bin and a rise of earth and a squat of clot of dark cloud. ... Who are we waiting for? I said. The dust said, We're last to arrive. And I looked again and saw our comrades; a tower block overhang, a copse of trees, sunk metal, water, a misshapen bin, the ground, vapor in the sky. Venue and participants were one.[45]

One might concur with Ann and Jeff VanderMeer that weird fiction is "darkly democratic"[46] precisely because it revels in a radically egalitarian allocation of voice and agency—the flat ontology of a discarded and commodified underclass in the planetary relations of production. Where venue and participants become one, the customary hierarchy between background and foreground is revoked. To describe grinding landmasses as the "subject-objects of history"[47] is thus to literalize "the crucial political task ... to distribute agency as far and in as differentiated a way as possible,"[48] which Latour views as the imperative of our time.

In contrast to Lovecraft, Miéville's vision of geological subjectivity implies new heterogeneous alliances, empowers the marginalized outcasts of industrial capitalism, and does not fold back onto an anthropocentric demarcation of inhuman alterity. Rather than raising the specter of destruction or an imperative to reinstate impermeable ontological boundaries, Miéville foregrounds transformation and portrays the entire planet as a congregation of political actors involved in diplomacy, agitation, radicalization, and, potentially, reconciliation. Mobilizing "The Dusty Hat" for an understanding of enmeshments between the human, the planet, and its other-than-human inhabitants means to acknowledge that, indeed, "venue and participants are one."

OF ODDKIN AND RELATIONS

The Anthropocene is a story, a convenient narrative that comes with its own preconceptions, focalizations, and intertexts. In imagining alternative visions for the future, "[i]t matters what stories tell stories,"[49] as Donna Haraway reminds her readers, just as much as it matters "what relations relate relations."[50] For her, the Anthropocene is a dangerously inadequate "tool, story, or epoch to think with,"[51] because as "a story of History that human exceptionalists tell" it "relies too much on what should be an 'unthinkable' theory of relations, namely the old one of bounded utilitarian individualism."[52] Rather than foregrounding ecological entanglement, it perpetuates the myth of an omnipotent (and, by default, white male) *anthropos*. Glossing over the alternative concept of the Capitalocene, which she recognizes as systemically more appropriate, yet too invested in "fundamentalist Marxism,"[53] she coins the "Chthulucene" as a more productive conceptual handle for envisioning heterogeneous relations in the current age. The homophonous reference to Lovecraft's Cthulhu mythos is meant not as an endorsement but as a bastardization of the tentacular Old Ones that have defined the trajectory of weird fiction since its inception. Her *chthulu* borrows from the Greek root *khthôn* (earth) and quite literally denotes geological subjectivities and subterranean forces:

> Chthonic ones are beings of the earth, both ancient and up-to-the-minute. I imagine chthonic ones as replete with tentacles, feelers, diggers, cords, whip-tails, spider legs, and very unruly hair. … Chthonic ones are monsters in the best sense; they demonstrate and perform the material meaningfulness of earth processes and critters. They also demonstrate and perform consequences. Chthonic ones are not safe; they have no truck with ideologues; they belong to no one; they writhe and luxuriate in manifold forms and manifold

names in all the airs, waters, and places of earth. They make and unmake; they are made and unmade.[54]

Haraway's Chthulucene is attuned to entanglements and transformations. Its defining mode of belonging and becoming is not the enforcement of ontological territoriality but "[m]aking kin as oddkin,"[55] which requires "unexpected collaborations and combinations" and supports a "material semiotics [that] is always situated, someplace and not noplace, entangled and worldly."[56] Building on Lynn Margulis's theory of symbiogenesis—the scientific infrastructure of the Gaia hypothesis—it implies "sympoiesis."[57]

In the remaining part of this chapter, I will propose Jemisin's *Broken Earth* trilogy (*The Fifth Season*, 2015; *The Obelisk Gate*, 2016; *The Stone Sky*, 2017) as a literary example of the type of estrangements of the weird that Haraway may have had in mind. Jemisin's novels decisively reject the Lovecraftian dehumanization of nonnormative subjectivities and link up with some of the potentially progressive politics of Miéville's fiction referenced above. Resonating with Jeff VanderMeer's description of the New Weird as "a type of secondary-world fiction that subverts the romanticized ideas about place found in traditional fantasy,"[58] Jemisin's story is set on an alter-Earth whose population lives in a condition of constant geo-climatological precarity. Chronicled in the trilogy's appendix, the history of this world unfolds as a millennia-long series of seismic catastrophes catalogued as "Fifth Seasons"[59] which trigger toxic winds, devastating environmental alterations, and prolonged winters. Echoing Haraway, the trilogy's title invokes "a vulnerable and wounded earth"[60] whose *ur*-trauma is the escalating industrial zeal by which "some arrogant, self-absorbed people tried to put a leash on the rusting planet"[61] and its ecologies: "They poisoned waters beyond even his ability to cleanse, and killed much of the other life that lived on his surface. They drilled through the crust of his skin, past the blood of his mantle to get to the sweet marrow of his bones."[62] The height of this hubris is an attempt to harness the very subterranean life-force of the planet—known as "magic." In an unspeakable attempt of resource extraction that goes disastrously wrong, the moon is flung out of its earthbound orbit leaving sentient Father Earth furious about the loss of his only child and unmollifiable in his determination to exact revenge leading to the first of the Fifth Seasons. Instrumentalized for this calamitous feat are a group of Frankensteinian hybrids—so-called tuners—who have been "orogenetically" enhanced to sense and manipulate the molecular forces suffusing all geological formations. They are inextricably attuned to the energetic web of the entire geosphere:

> All energy is the same, through its different states and names. Movement creates heat which is also light that waves like sound which tightens or loosens the atomic bonds of crystal as the hum with strong and weak forces. In mirroring resonances with all of this is magic, the radiant emission of life and death. This is our role: To weave together those disparate energies. To manipulate and mitigate and, through the prism of our awareness, produce a singular force that cannot be denied. To make of cacophony, symphony.[63]

It is instructive to recall that Haraway heralds not Lovecraft but a spider named after his most notorious tentacled demon as the champion of her Chthulucene—*Pimoa cthulhu*: "I remember that *tentacle* comes from the Latin *tentaculum*, meaning 'feeler,' and *tentare*, meaning 'to feel' and 'to try'; and I know that my leggy spider has many-armed allies. Myriad tentacles will be needed to tell the story of the Chthulucene."[64] Akin to Haraway's spider, Jemisin's heroes are weavers whose engagement with the energetic flows of the planet follows a tactile economy of touch and vibration. As one of them notes about nonsensitive humans, "hair textures and eye shapes and unique body odors each have as much meaning to them as the perturbations of tectonic plates have to me."[65] Their names are unutterable and indicative of the geomorphic nature of their subjectivities: "*deep stab, breach of clay sweetburst, soft silicate underlayer, reverberation,*"[66] "*cracked geode taste of adularescent salts, fading echo.*"[67] Also deriving from their planetary connection is the ability to communicate in "earth-talk,"[68] allowing them to supplement audible speech with nuance via the surrounding mineral medium, signaling "crumble of resentment,"[69] "an acid, pressurized boil of bitterness,"[70] or "a pent, angry jitter of the entire ambient, air molecules shivering."[71] Similar to Miéville, Jemisin persistently employs literary registers that align human and geological affect, and readers will likely detect rapports with the idiom of contemporary materialist ecocriticism, whether with regard to Jane Bennett's "material vibrancy,"[72] Cohen's "human-lithic enmeshment,"[73] or Kathryn Yusoff's "geologic life."[74]

What the tuners soon realize is that they are perceived as a threat by the human powers that be: "With every glimpse of normalcy the city teaches us just how abnormal we are."[75] As one of them puts it, "monsters like us were the enemy of all good people,"[76] In what can readily be received as a commentary on the complicity between resource capitalism and a liberal humanist project historically engaged in the rigorous policing of legitimate forms of personhood, the tuners are both tools and expendable.

Jemisin focuses on the fate of the tuners' dehumanized evolutionary descendants—the caste of so-called orogenes, derogatively referred to as "roggas"—whose powers are honed, exploited, and supervised at the Fulcrum—nominally a training academy, but effectively a paramilitary ghetto designed to produce docile bodies and shield the rest of the population from the specter of orogenic monstrosity. As one of its founding legal documents outlines: "though they must be managed with kind hands to the benefit of both bond and free, any degree of orogenic capability must be assumed to negate its corresponding personhood. They are rightfully regarded as an inferior and dependent species."[77] Readers follow the journey of Essun, a particularly gifted orogene destined to mend the relationship between people and planet by returning the moon to its orbit and in the process rehabilitate her kin.

Perhaps most fundamentally, the *Broken Earth* trilogy tells a story of troubled relations and odd kinship—of the trauma of displacement and exclusion and the fragile vision of ecological reattachment, and collaborative action. The plot opens to a raging Earth whose child was stolen, a mother whose son is killed and who embarks on a search for her missing daughter, a human civilization estranged from its volatile planetary habitat, a society militantly stratified and committed to the enslavement and erasure of illegitimate hybrids, and an ecological condition in which networks of "magic" are atrophied and withered. All of these conditions are the cause or result of devastating eruptions of violence. The original sin of casting out the moon is perpetrated by Hoa, a tuner who in an ultimate act of anticolonial resistance decides to incite the planet's ire and direct it at all life on Earth in order to see the civilization of his oppressors destroyed. The bitterness in his justification is palpable: "Well, some worlds are built on fault lines of pain, held up by nightmares. Don't lament when those worlds fall. Rage that they were built doomed in the first place."[78] Several millennia and countless cataclysmic Fifth Seasons later, the powerful orogene Alabaster uses his geomorphic gift to annihilate the heart of the rebuilt human empire and with it the carceral institution that has held his people captive: In a dramatic gesture of determination,

> he reaches deep and takes hold of the humming tapping bustling reverberating rippling vastness of the city, and the quieter bedrock beneath it, and

the roiling churn of heat and pressure beneath that. ... He takes all that, the strata and the magma and the people and the power, in his imaginary hands. Everything. He holds it. He is not alone. The earth is with him. Then he breaks it.[79]

Geology provides the presiding source of metaphors. The parallelism between geological and psychological states—with fault lines, cracks, and fractures—is programmatic in that readers and characters are increasingly uncertain whether "the distinction between living creature and lifeless object matters."[80] The permeability of this boundary is indicative of the modes of kinship and transcorporeality that ultimately serve to heal Jemisin's wounded subjectivities. Characters are not stable but—like the planet—evolve in a process of perpetual transformation. Tuners and orogenes are reconstituted as stone-eaters—enigmatic immortal beings who look like marble statues, travel through solid rock with lightening speed and ease, and engage in symbiotic relationships with powerful orogenes, feeding off their petrified body parts sacrificed in particularly forceful feats of geomorphic manipulation. Essun herself is known by a variety of names—Damaya, Syenite—that attach to different stages of her biography. As the narrator—who turns out to be Hoa—imparts to Essun, identity is relational and multilaterally emplaced; everyone contains multitudes:

> After all, a person is herself, and others. Relationships chisel the final shape of one's being. I am me, and you. Damaya was herself *and* the family that rejected her *and* the people of the Fulcrum who chiseled her to a fine point. Syenite was Alabaster *and* Innon *and* the people of poor lost Allia and Meov. Now you are Tirimo *and* the ash-strewn road's walkers *and* your dead children...and also the living one that remains.[81]

Literalized here, we find what Haraway calls becoming-with, a celebration of oddkinship among monstrous outcasts and chthonic beings uncontainable by myths of bounded personhood. Joining a wandering heterogeneous band of humans, orogenes, so-called guardians, and stone-eaters, Essun and her equally powerful daughter Nassun learn that survival is contingent on collaboration, trust, and reciprocal response-abilities. About eleven-year-old Nassun Hoa notes, "she will learn the word *symbiosis* and nod, pleased to have a name for it at last. But long before that, she will have already decided that *family* will do."[82] At the trilogy's climax, Essun and her daughter face off in a struggle over two competing strategies on how to resolve the injustices they were forced to endure—cataclysmic retaliation or hope in coexistence. Deeply affected by her mother's eventual self-sacrifice, Nassun aborts her original plan to utilize the planet's energy and transform every living being into stone, and instead settles for guiding the moon back into an earthbound orbit in hope of appeasing Father Earth, achieving a truce, and finally ending the Fifth Seasons. Charged with the utopian promise of a reconciliation with the planet, this ending, however, leaves doubts whether the colonial trauma that underpins Jemisin's story will truly be healed.

TOWARD GEOLOGICAL SURVIVANCE

Faced with the challenges presented by the recognition of geologic life, Yusoff asks, "How do we speak of deep time and inhuman beginnings within the context of these Earth forces in ways that offer a generative politics of minerality, rather than one of unilateral destruction?"[83] Haraway makes a corresponding call: "How can we think in times of urgencies *without* the self-indulgent and self-fulfilling myths of apocalypse, when every fiber of our being is interlaced, even complicit, in the webs of processes that must somehow be engaged and repatterned?"[84] It seems that these are the types of questions that the *Broken Earth* trilogy invites us to engage with. Confronted with ubiquitous invocations of extinction, undeniable environmental devastation, and traumatic histories of eco- and genocide, how can we invigorate visions of non-anthropocentric survival that anchor in more egalitarian recognitions of voice and agency?

Quite explicitly, Jemisin's work was written in response to ongoing histories of systemic racism and inspired by the Black Lives Matter movement. Readers have, moreover, perceptively commented on the parallels between the internment of orogenes at the Fulcrum and the systematic removal of Indigenous children from their families and subsequent confinement in residential schools in Canada and Australia.[85] The treatment of orogenes is frequently invoked in terms of both literal and cultural genocide:

> If every orogene is hunted down and slain, and if the neck of every orogene infant born thereafter is wrung, and if every one like me who carries the trait is killed or effectively sterilized, and if even the notion that orogenes are human is denied…that would be genocide. Killing a people down to the very *idea* of them as people.[86]

The trauma that haunts Jemisin's narrative is both geomorphic *and* colonial—a historical imbrication compellingly unfolded in Yusoff's recent study *A Billion Black Anthropocenes or None* (2019). Acknowledging Jemisin's poetics as a decisive influence, Yusoff makes an important intervention in Anthropocene scholarship by illustrating how "extractive geo-logics"[87] have facilitated and perpetuated the categorization of racialized populations as inhuman, emphasizing that "any critical theory [of the Anthropocene] that does not work with and alongside black and indigenous studies ... will fail to deliver any epochal shift at all."[88] Following Yusoff, I suggest leaving the usual Eurocentric materialist thinkers of the Anthropocene aside and seeking out the affinities between the *Broken Earth* trilogy and decolonial scholarship by contemporary Indigenous critics. It is in this vein, that we may read the struggle of Jemisin's orogenes through the lens of what Anishinaabe writer Gerald Vizenor has called *survivance*—frequently interpreted as a portmanteau of "survival" and "resistance," or "insurgence."[89] Central to self-determined practices of survivance as a strategy of resisting cultural genocide—the systematic effort to erase a people's cultural identity experienced by Indigenous populations in settler-colonial states worldwide—is the folding of storytelling, emplacement, and embodied relationships.[90] As Cherokee critic and author Daniel Heath Justice explains, "relationships are storied, imagined things; they set the scope for our experience of being and belonging."[91] Writing about the significance of storytelling in Indigenous literatures, he continues:

> [S]tory makes meaning of the relationships that define who we are and what our place is in the world; it reminds us of our duties, our rights and responsibilities, and the consequences and transformative possibilities of our actions. It also highlights what we lose when those relationships are broken or denied to us, and what we might gain from even partial remembrance.[92]

In many Indigenous frameworks, kinship, relations, land, and stories are understood as reciprocal synechdoches—they are mutually constitutive. Beyond clichéd notions of the "ecological Indian,"[93] it may not surprise, in this regard, that the concept of a sentient planet so evocatively explored by Jemisin is integral to many Indigenous frameworks.[94] Exemplary is Haudenosaunee and Anishnaabee scholar Vanessa Watts' introduction of "Place-Thought," which "is based upon the premise that land is alive and thinking and that humans and non-humans derive agency through the extensions of these thoughts."[95]

Recognizing how the connection of storytelling and kinship becomes an integral function of Anthropocene survival for the colonized subjects in Jemisin's novels may begin by noting that the narrative is presented as a type of frame tale through which Hoa recounts the world's events to Essun so that after her death as orogenic human and her ensuing reassemblage as a stone-eater she will retain as much of her identity as possible. It seems also fitting that one of the allies that offers to accompany Essun on her final most dangerous journey is a Lorist, a storyteller or chronicler who embraces her task as "warning of the coming holocaust and teaching others how to cooperate, adapt, and remember."[96] Acknowledging stories themselves as lived relationships is part and parcel of the broadened notion of oddkinship emphasized by Haraway and entails that "the range of relatives to whom we are responsible," as Justice notes, "extends far beyond our biological relatives and, indeed, the category of the human itself."[97] All earth beings in Jemisin's trilogy derive their life-force from the lines of "magic" or "silver" that connect them to one another and the planet's own geo-energetic web of relations. As Nassun realizes, "[t]his silver deep within Father Earth wends between the mountainous fragments of his substance in exactly the same way that they twine among the cells of a living, breathing thing. And that is because *a planet* is a living, breathing thing...."[98]

If, indeed, Jemisin metabolizes the Anthropocene through her poetic portrayal of planetary–human relationships, she amplifies models of material relations and agency that actively address and resist the (often neglected) ideologies of extractive colonialism that have shaped its foundations. In resonance with Yusoff, Jemisin champions "histories that launch a praxis for an insurgent geology."[99] If this counts as "weird," then this weirdness derives not from the repressed alterity of an inhuman underground but from the radical alliances that are bound to emerge when colonized peoples and colonized matter are empowered as kindred respondents to the crisis commonly referred to as the Anthropocene.

NOTES

1. See Paul J. Crutzen, "Geology of Mankind: The Anthropocene," *Nature* 415 (2002): 23; and John Harrison et al., "New Weird Discussions: The Creation of a Term," in *The New Weird*, eds. Jeff VanderMeer and Ann VanderMeer (San Francisco: Tachyon Publications, 2008), 317–31.

2. I here rely on Jeffrey Jerome Cohen, who draws on the concept of the "'the inhuman' to emphasize both difference ('in-' as in negative prefix) and intimacy ('in-' as indicator of estranged interiority)" (10).
3. Bruno Latour, "Agency at the Time of the Anthropocene," *New Literary History* 45 (2014): 4.
4. Eugene Thacker, *In the Dust of This Planet* (Winchester: Zero Books, 2011), 4.
5. Ibid., 6.
6. See Latour, "Agency at the Time of the Anthropocene," 3.
7. See Stubby the Rocket, "N. K. Jemisin's New Contemporary Fantasy Trilogy Will Mess with the Lovecraft Legacy." Tor.com, August 18, 2017. Accessed March 14, 2019, https://www.tor.com/2017/08/18/nk-jemisin-lovecraft-trilogy/.
   Her upcoming trilogy is announced as an explicit reckoning with Lovecraft's toxic legacy. See Lila Shapiro. "For Reigning Fantasy Queen N.K. Jemisin, There's No Escape from Reality." Vulture, November 29, 2018. Accessed March 14, 2019, https://www.vulture.com/2018/11/nk-jemisin-fifth-season-broken-earth-trilogy.html.
8. See Shapiro, "For Reigning Fantasy Queen N.K. Jemisin, There's No Escape from Reality."
9. See, for instance, Gerald Vizenor, "Postindian Warriors," in *Learn, Teach Challenge: Approaching Indigenous Literatures*, eds. Deanna Reder and Linda M. Morra (Waterloo, ON: Wilfred Laurier University Press, 2016), 155168.
10. Kathryn Yusoff, *A Billion Black Anthropocenes or None* (Minneapolis: University of Minnesota Press, 2019), xiii.
11. Stephen Jay Gould, *Time's Arrow, Time's Cycle: Myth and Metaphor in the Discovery of Geological Time* (Cambridge, MA: Harvard University Press, 1997), 2.
12. H.P. Lovecraft, "The Shadow Out of Time," in *Necronomicon: The Best Weird Tales of H.P. Lovecraft*, ed. Stephen Jones (London: Gollancz, 2008), 603.
13. Ibid., 568.
14. Ibid., 579.
15. Ibid., 593.
16. Prime examples of the continuing allure of geological insurrections in Lovecraft's wake can be found in Robert Barbour Johnson's "Far Below" (1939) and Fritz Leiber's "The Black Gondolier" (1964).
17. Elaine L. Graham, *Representations of the Post/Human: Monsters, Aliens and Others in Popular Culture* (Manchester: Manchester University Press, 2002), 12.
18. Shapiro, "For Reigning Fantasy Queen N.K. Jemisin, There's No Escape from Reality."

19. Michael Moorcock. "Foreweird," in *The Weird: A Compendium of Strange and Dark Tales*, eds. Ann VanderMeer and Jeff VanderMeer (New York: Tom Doherty Associates, 2011), xiii.

20. Donna J. Haraway, *Staying with the Trouble: Making Kin in the Chthulhucene.* (Durham and London: Duke University Press, 2016), 12.

21. Ibid.

22. China Miéville, "Reveling in Genre: An Interview with China Miéville," interview by Joan Gordon, *Science Fiction Studies* 30, no. 3 (2003). Accessed October 4, 2018, http://www.depauw.edu/sfs/interviews/Miévilleinterview.htm.

23. See, for instance, China Miéville, "Covehithe," in *Three Moments of an Explosion: Stories* (New York: Del Rey, 2015), 299–311; and "Polynia," in *Three Moments of an Explosion: Stories* (New York: Del Rey, 2015), 5–22.

24. China Miéville, "The Dusty Hat," in *Three Moments of an Explosion: Stories* (New York: Del Rey, 2015), 200.

25. Ibid., 208.

26. Ibid.

27. Miéville, "The Dusty Hat," 213.

28. Ibid., 207.

29. Ibid.

30. Ibid.

31. Ibid., 209.

32. Ibid.

33. Ibid.

34. Ibid.

35. Ibid., 211–12.

36. Ibid., 212.

37. Ibid.

38. Ibid.

39. Ibid.

40. Ibid.

41. China Miéville, "Afterweird: The Efficacy of a Worm-Eaten Dictionary," in *The Weird: A Compendium of Strange and Dark Tales*, eds. Ann VanderMeer and Jeff VanderMeer (New York: Tom Doherty Associates, 2011), 1115.

42. Miéville, "The Dusty Hat," 212.

43. Ibid., 214.

44. Bruno Latour, *We Have Never Been Modern* (Cambridge: Harvard University Press, 1993), 124.

45. Miéville, "The Dusty Hat," 216.

46. Ann VanderMeer and Jeff VanderMeer, "Introduction," in *The Weird: A Compendium of Strange and Dark Tales*, eds. Ann VanderMeer and Jeff VanderMeer (New York: Tom Doherty Associates, 2011), xvi.

47. Miéville, "The Dusty Hat," 212.

48. Latour, "Agency at the Time of the Anthropocene," 17.
49. Haraway, *Staying with the Trouble*, 35.
50. Ibid.
51. Ibid., 49.
52. Ibid.
53. Ibid., 50.
54. Ibid., 2.
55. Ibid.
56. Ibid., 4.
57. Ibid., 58.
58. Jeff VanderMeer, "Introduction—The New Weird: 'It's Alive?'," in *The New Weird*, eds. Jeff VanderMeer and Ann VanderMeer (San Francisco: Tachyon Publications, 2008), xvi.
59. N.K. Jemisin, *The Fifth Season: The Broken Earth Trilogy I* (London: Orbit Books, 2015), 451.
60. Haraway, *Staying with the Trouble*, 10.
61. N.K. Jemisin, *The Stone Sky: The Broken Earth Trilogy III* (London: Orbit Books, 2017), 313.
62. Jemisin, *The Fifth Season*, 379–80.
63. Jemisin, *The Stone Sky*, 97.
64. Haraway, *Staying with the Trouble*, 31.
65. Ibid., 44.
66. Jemisin, *The Stone Sky*, 44; italics in original.
67. Ibid., 46; italics in original.
68. Ibid., 47.
69. Ibid., 100.
70. Ibid., 213.
71. Ibid., 101.
72. Jane Bennett, *Vibrant Matter: A Political Economy of Things* (Durham and London: Duke University Press, 2010), xiii.
73. Cohen, *Stone: An Ecology of the Inhuman*, 6.
74. Kathryn Yusoff, "Geologic Life: Prehistory, Climate, Futures in the Anthropocene," *Environment and Planning D: Society and Space* 31 (2013).
75. Ibid., 204.
76. Jemisin, *The Stone Sky*, 203.
77. N.K. Jemisin, *The Obelisk Gate: The Broken Earth Trilogy II* (London: Orbit Books, 2016), 258.
78. Jemisin, *The Stone Sky*, 7.
79. Jemisin, *The Fifth Season*, 7.
80. Jemisin, *The Stone Sky*, 235.
81. Jemisin. The Obelisk Gate, 1.
82. Jemisin, *The Stone Sky*, 132.

83. Yusoff, "Geologic Life: Prehistory, Climate, Futures in the Anthropocene," 780.
84. Haraway, *Staying with the Trouble*, 35.
85. See John Joseph Adams and David Barr, "Black Lives Matter Inspired This Chilling Fantasy Novel." *Wired*, August 29, 2015. Accessed March 14, 2019, https://www.wired.com/2015/08/geeks-guide-nk-jemisin/.
86. Jemisin, *The Stone Sky*, 117.
87. Yusoff, *A Billion Black Anthropocenes or None*, 16.
88. Ibid., 18.
89. See, for instance, Gerald Vizenor, "Postindian Warriors," in *Learn, Teach Challenge: Approaching Indigenous Literatures*, eds. Deanna Reder and Linda M. Morra. (Waterloo, ON: Wilfred Laurier University Press, 2016), 155–68.
90. Also see Leanne Betasamosake Simpson, "Theorizing Resurgence from Within Nishnaabeg Thought," in *Dancing on Our Turtle's Back* (Winnipeg: ARP Books, 2011), 31–47.
91. Daniel Heath Justice, *Why Indigenous Literatures Matter* (Waterloo, ON: Wilfred Laurier University Press, 2018), 74.
92. Ibid., 75.
93. See Shepard Krech III, *The Ecological Indian: Myth and History* (New York: W. W. Norton, 1999).
94. See, for example, Marisol de la Cadena's *Earth Beings: Ecologies of Practice in the Andean Worlds* (2015), an anthropological account of the imbrications among Indigenous liberation movements and the recognition of mountains, animals, and rivers as sentient "earth-beings."
95. Vanessa Watts, "Indigenous Place-Thought & Agency Amongst Humans and Non-humans (First Woman and Sky Woman Go on a European World Tour!)," *Decolonization: Indigeneity, Education & Society* 2, no.1 (2013): 21.
96. Jemisin, *The Stone Sky*, 343.
97. Justice, *Why Indigenous Literatures Matter*, 75.
98. Jemisin, *The Stone Sky*, 242.
99. Yusoff, *A Billion Black Anthropocenes or None*, 22.

Bibliography

Adams, Joseph, and David Barr. "Black Lives Matter Inspired This Chilling Fantasy Novel." *Wired*, August 29, 2015. Accessed May 15, 2019. https://www.wired.com/2015/08/geeks-guide-nk-jemisin/.
Bennett, Jane. *Vibrant Matter: A Political Ecology of Things*. Durham and London: Duke University Press, 2010.
Canavan, Gerry, and Andrew Hageman. *Global Weirding*. Vashon: Paradoxa, 2016.

Cohen, Jeffrey Jerome. *Stone: An Ecology of the Inhuman*. Minneapolis: University of Minnesota Press, 2015.

Crutzen, Paul J. "Geology of Mankind: The Anthropocene." *Nature* 415 (January 2002): 23.

De La Cadena, Marisol. *Earth Beings: Ecologies of Practices Across the World*. Durham and London: Duke University Press, 2015.

Gould, Stephen Jay. *Time's Arrow, Time's Cycle: Myth and Metaphor in the Discovery of Geological Time*. The Jerusalem-Harvard Lectures. Cambridge, MA: Harvard University Press, 1987.

Graham, Elaine. *Representations of the Post/Human: Monsters, Aliens and Others in Popular Culture*. Manchester: Manchester University Press, 2002.

Haraway, Donna Jeanne. *Staying with the Trouble: Making Kin in the Chthulucene*. Experimental Futures: Technological Lives, Scientific Arts, Anthropological Voices. Durham: Duke University Press, 2016.

Harrison, M. John. "New Weird Discussions: The Creation of a Term." In *The New Weird*, edited by Jeff VanderMeer and Ann VanderMeer, 317–31. San Francisco: Tachyon Publications, 2008.

Jemisin, N.K. *The Fifth Season*. London: Orbit, 2015.

———. *The Obelisk Gate*. London: Orbit, 2016.

———. *The Stone Sky*. London: Orbit, 2017.

Johnson, Robert Barbour. "Far Below." In *The Weird: A Compendium of Strange and Dark Stories*, edited by Ann VanderMeer and Jeff VanderMeer, 260–67. New York: Tom Doherty Associates, 2011.

Justice, Daniel Heath. *Why Indigenous Literatures Matter*. Waterloo, ON: Wilfrid Laurier University Press, 2018.

Krech III, Shepard. *The Ecological Indian: Myth and History*. New York: W. W. Norton, 1999.

Latour, Bruno. "Agency at the Time of the Anthropocene." *New Literary History* 45 (2014): 1–18.

———. *We Have Never Been Modern*. Translated by Catherine Porter. Cambridge: Harvard University Press, 1993.

Leiber, Fritz. *The Black Gondolier and Other Stories*. New York: Open Road Media, 2014.

Lovecraft, H.P. "The Shadow Out of Time." In *Necronomicon: The Best Weird Tales of H.P. Lovecraft*, edited by Stephen Jones, 555–606. London: Gollancz, 2008.

Miéville, China. "Afterweird: The Efficacy of a Worm-Eaten Dictionary." In *The Weird: A Compendium of Strange and Dark Tales*, edited by Ann VanderMeer and Jeff VanderMeer, 1113–16. New York: Tom Doherty Associates, 2011.

———. "Covehithe." In *Three Moments of an Explosion: Stories*, 1st ed., 299–311. New York: Del Rey, 2015.

———. "Polynia." In *Three Moments of an Explosion: Stories*, 1st ed., 5–22. New York: Del Rey, 2015.

———. "Reveling in Genre: An Interview with China Miéville." Interview by Joan Gordon. *Science Fiction Studies* 30, no. 3, November 2003. http://www.depauw.edu/sfs/interviews/Miévilleinterview.htm.

———. "The Dusty Hat." In *Three Moments of an Explosion: Stories*, 1st ed., 197–218. New York: Del Rey, 2015.

Moorcock, Michael. "Foreweird." In *The Weird: A Compendium of Strange and Dark Tales*, edited by Ann VanderMeer and Jeff VanderMeer, xi–xiv. New York: Tom Doherty Associates, 2011.

Shapiro, Lila. "For Reigning Fantasy Queen N.K. Jemisin, There's No Escape from Reality." Vulture, November 29, 2018. https://www.vulture.com/2018/11/nk-jemisin-fifth-season-broken-earth-trilogy.html.

Simpson, Leanne Betasamosake. "Theorizing Resurgence from Within Nishnaabeg Thought." In *Dancing On Our Turtle's Back*, 31–47. Winnipeg: ARP Books, 2011.

Stubby the Rocket. "N. K. Jemisin's New Contemporary Fantasy Trilogy Will Mess with the Lovecraft Legacy." Tor.com, August 18, 2017. https://www.tor.com/2017/08/18/nk-jemisin-lovecraft-trilogy/.

Thacker, Eugene. *In the Dust of This Planet.* 1. publ. Horror of Philosophy 1. Winchester: Zero Books, 2011.

Tsing, Anna Lowenhaupt, ed. *Arts of Living on a Damaged Planet: Ghosts and Monsters of the Anthropocene.* Minneapolis: University of Minnesota Press, 2017.

VanderMeer, Ann and Jeff VanderMeer. "Introduction." In *The Weird: A Compendium of Strange and Dark Tales*, edited by Ann VanderMeer and Jeff VanderMeer, xv–xx. New York: Tom Doherty Associates, 2011.

VanderMeer, Jeff. "Introduction—The New Weird: 'It's Alive?'" In *The New Weird*, edited by Jeff VanderMeer and Ann VanderMeer, ix–xviii. San Francisco: Tachyon Publications, 2008.

Vizenor, Gerald. "Postindian Warriors." In *Learn, Teach Challenge: Approaching Indigenous Literatures*, edited by Deanna Reder and Linda M. Morra, 155–68. Waterloo, ON: Wilfred Laurier University Press, 2016.

Watts, Vanessa. "Indigenous Place-Thought & Agency Amongst Humans and Non-humans (First Woman and Sky Woman Go on a European World Tour!)." *Decolonization: Indigeneity, Education & Society* 2, no. 1 (2013): 20–34.

Yusoff, Kathryn. *A Billion Black Anthropocenes or None.* Minneapolis: University of Minnesota Press, 2019.

———. "Geologic Life: Prehistory, Climate, Futures in the Anthropocene." *Environment and Planning D: Society and Space* 31 (2013): 779–95.

CHAPTER 7

"Indifference Would Be Such a Relief": Race and Weird Geography in Victor LaValle and Matt Ruff's Dialogues with H. P. Lovecraft

James Kneale

Remapping Lovecraft Country

This chapter considers the weird geographies presented in two novels published in 2016, Victor LaValle's *Ballad of Black Tom* and Matt Ruff's *Lovecraft Country*. LaValle's novella tells the story of Tommy Tester, a young black man hustling for a living in New York in 1925. Tommy gets caught up in the occult schemes of Robert Suydam, takes on a new name—Black Tom—and attracts the attention of a police detective named Malone. The story is a revision of one of H. P. Lovecraft's most unpleasant and incoherent stories, "The Horror at Red Hook." In LaValle's version, Suydam's cult wishes to end the world by waking the Sleeping King (Cthulhu). LaValle dedicates *Ballad* to Lovecraft ("with all my conflicted feelings"), and Lovecraft and his wife Sonia Greene feature, unnamed, in the novella.[1]

J. Kneale (✉)
Department of Geography, University College, London, UK

© The Author(s) 2019

J. Greve and F. Zappe (eds.), *Spaces and Fictions of the Weird and the Fantastic*, Geocriticism and Spatial Literary Studies, https://doi.org/10.1007/978-3-030-28116-8_7

Ruff's *Lovecraft Country* concerns an African-American family and their friends living in Chicago in 1954. George Berry is the editor and publisher of *The Safe Negro Guide*, based on the real *Negro Motorist Green Book*, created "to give the Negro traveller information that will keep him from running into difficulties, embarrassments and to make his trips more enjoyable."[2] His half-brother, Montrose Turner, is still scarred by his father's death in the Tulsa Massacre of 1921; Montrose's son Atticus, a veteran of the Korean War, works for George, as does George's wife, Hippolyta. The Berrys, the Turners and their friends get embroiled in a struggle between rival white sorcerers. Some of Ruff's characters still love Lovecraft's stories though, as African-Americans, they are clearly not his intended readers. George tells his nephew "stories are like people, Atticus. Loving them doesn't make them perfect. You try to cherish their virtues and overlook their flaws. The flaws are still there, though."[3]

There have been many other authors who have written in dialogue with Lovecraft, from his original circle to the ubiquitous references in contemporary popular culture, but the fictions that are the focus of this chapter were written to challenge Lovecraft's unabashed and extreme racism.[4] Earlier authors rarely engaged with this directly, with a few significant exceptions.[5] In 2016, however, LaValle's and Ruff's novels were published alongside Kij Johnson's *The Dream-Quest of Vellitt Boe* and Brian M. Sammons and Oscar Rios' collection *Heroes of Red Hook*.[6] All four challenge the racial and/or sexual politics of Lovecraft's stories, though LaValle and Ruff's fictions focus on his racism.

These novels are highly conscious of their intertextuality. In *Ballad*, Black Tom growls "I bear a hell within me … And finding myself unsympathized with, wished to tear up the trees, spread havoc and destruction around me, and then have sat down and enjoyed the ruin."[7] Here Lavalle echoes Mary Shelley's *Frankenstein*; like Frankenstein's creation, Tom experiences rejection and desires revenge. Similarly, in *Lovecraft Country*, Ruby Dandridge, transformed into a white woman, reads Robert Louis Stevenson's *Strange Case of Jekyll and Hyde*. She notes the racialization of morality in the contrast between Jekyll's hand, "large, firm, white, and comely," and Hyde's, which is "lean, corded, knuckly, of a dusky pallor and thickly shaded with a swart growth of hair."[8] But the most important intertextual dialogues are, of course, with Lovecraft.

In 1915 Lovecraft argued that the First World War represented "the supremest of all crimes… the violation of race" because the "Teutonic races" fought one another rather than uniting to "crush successively the rising power of Slav and Mongolian." In another essay published that year, he described an article by a (Jewish) writer which criticized D. W. Griffith's *The Birth of a Nation* as "a crime which in a native American of Aryan blood would be deserving of severe legal punishment"; he then defended the Ku Klux Klan. He offered only slightly qualified support for Hitler as late as 1933.[9] Lovecraft was horrified by miscegenation and threats to racial purity.[10] He hated change, often seeing it as a racialised threat to an imagined past. He found "many highly picturesque ... reliques of better days" in Flatbush, for example, though "these were wholly surrounded by the incursions of decadent modernity"—which he blamed on Jewish investment.[11] His racism was polemical and defensive; as Michel Houellebecq suggests, Lovecraft "designated himself the victim and ... picked his tormentors."[12]

Those critics who have addressed Lovecraft's racism have tended to follow Houellebecq's argument that "Racial hatred provokes in Lovecraft the trancelike poetic state" in which he piles up the adjectives for which he is famous.[13] Mark Fisher argued that Lovecraft's hatred "transforms an ordinary object causing displeasure into a Thing which is both terrible and alluring, which can no longer be libidinally classified as either positive or negative."[14] As China Miéville points out, we might have to confront the fact that Lovecraft's writing is effective not despite his racism, but *because* of it.[15]

This chapter will argue that these recent novels by LaValle and Ruff make it possible to think of white racism as itself weird or eerie, in Lovecraft's day or our present. After reviewing recent work on the weird and exploring some of its geographies, the chapter returns race to these discussions, before examining the weird and eerie spaces presented by the two novels. It concludes by suggesting that our reading of the weird might be expanded to make room for other kinds of horror.

Weird Geographies

There has been a good deal of interest in the weird in recent years.[16] In part, this reflects the critical reevaluation of authors like Lovecraft and Arthur Machen, the revival of interest in figures like Leonora Carrington and Vernon Lee, and the boom in "New Weird" fiction written by M. John Harrison, China Miéville, Jeff VanderMeer, and others. At the same time, a number of philosophers have taken Lovecraft and the weird as the starting points for an exploration of pessimism and nihilism, and the weirdness of reality itself.[17]

Weird fiction seems well suited to the development of literary-geographical questions. As Fisher points out, "the notion of *the between* is crucial to the weird."[18] On the one hand, this reminds us that these fictions concern realms outside our own, leaking into or interpenetrating the human world.[19] This outside is often a nonhuman world, where human values are meaningless. Eugene Thacker suggests that horror and science fiction can help us to try to think about "the world-without-us"—the world we are left with after "the subtraction of the human from the world."[20] This aspect of weird space is closely bound up with arguments about Lovecraftian cosmic horror, nihilism, and pessimism.

On the other hand, the weird "between" encourages thoughts of relation or connection, proximity, and distance. Graham Harman argues that Lovecraft's allusive style draws our attention to the *gaps* "between objects and the power of language to describe them, or between objects and the qualities they possess."[21] Timothy Morton, like Harman, is interested in the weird relations that objects have with each other; Jane Bennett's "new materialism" makes room for lively, enchanting nonhuman matter; and Donna Haraway's "Chthulucene" helps us think through the "diverse earth-wide tentacular powers and forces and collected things" that make up posthuman worlds.[22] Networks, meshes, webs: the weird is made up of connections.

At the same time geographers have insisted that space, too, is relational. Places are formed out of the relations between things, and from their relations with other places. As Doreen Massey says, "what gives a place its specificity is not some long internalized history but the fact that it is constructed out of a particular constellation of relations, articulated together at a particular locus." For Haraway, "Nothing is connected to everything; everything is connected to something."[23] This is not the standardized, placeless world of 'globalization'; this is a world of shifting, uneven, complex relations. Morton argues that "Place now has nothing to do with good old reliable constancy. What has dissolved is the idea of *constant presence*: the myth that something is real insofar as it is consistently, constantly 'there'."[24] The relational character of place can be deeply weird in its own right. If haunting is, in its simplest form, a question of absent presence, then the traces of distant times and places haunt every site. If we add in Latour's "missing masses," the nonhumans normally absent from our understandings of place, places become even livelier. Morton's "hyperobjects"—like global warming—leave uncanny footprints in local places.

It is relatively easy to read Lovecraft's geographies in this way, as his professed "antiquarianism & exoticism" drove him to seek the past and the distant in the present and familiar.[25] His fictional places were defined by worldly connections, as Evan Lampe demonstrates.[26] Rhode Island ships were the most significant American carriers of enslaved people before and after the Revolution. Providence was shaped by its asymmetrical relations with West Africa, the Caribbean, Europe, and the other Atlantic colonies, entangled in Caribbean molasses, New England rum and cotton, and European buttons—and, crucially, the ships that brought these places, people and things together.[27]

But what makes these geographies *weird*? Harman identifies a passage in Lovecraft's "The Whisperer in Darkness" as a model for weird geography; it suggests that the distribution of Vermont's hamlets and roads was shaped by occult forces, as the locals avoided hills and valleys haunted by otherworldly entities.[28] We might look, then, for unexpected traces of agency in place, since agency implies relations between agents, and relations are spatial. Fisher's comparison of the weird and the eerie seems to support this argument: "the weird is constituted by a presence—the presence of *that which does not belong*." The eerie, on the other hand, is "is constituted by a *failure of absence* or by a *failure of presence*." The eerie cry that seems to come from nowhere exemplifies the first failure; the second can be found in "landscapes partially emptied of the human." In both cases "the eerie is fundamentally tied up with questions of agency." This eerie agency can be nonhuman, for example, "the agency of minerals and landscape for authors like Nigel Kneale and Alan Garner... 'we' 'ourselves' are caught up in the rhythms, pulsions and patternings of non-human forces."[29] The presence or absence of human or nonhuman agency in place makes weird or eerie geographies.

"Either Englishmen or Nothing Whatever"

These investigations of the weird do not share Lovecraft's racist politics—not least because a genuine posthumanism should be properly skeptical of ideas of "race." In fact, Lovecraft knew very well that his attachment to his corner of New England made little sense in terms of the nihilism and cosmic fear that he saw in the universe. As he argued in 1930:

> It is *because* the cosmos is meaningless that we must secure our individual illusions of values, direction, and interest by upholding the artificial streams which gave us such worlds of salutary illusion. That is—since nothing means anything in itself, we must preserve the proximate and arbitrary background which makes things around us seem as if they did mean something. In other words, we are either Englishmen or nothing whatever.[30]

Confronted with this choice, Lovecraft chose 'Englishness' over cosmic pessimism, even though he knew the former to be meaningless and arbitrary. Lovecraft clung to conservative, familiar and local constructions of nation, self and—above all—race while proclaiming the meaninglessness of human standards in the face of an uncaring universe.

If Lovecraft was not entirely honest about the universality of the human subject he set against the cosmos, then perhaps we should also scrutinize the subject of the weird thinking explored above. Fisher suggests that our encounters with nonhuman agency show us that "[t]here is no inside except as a folding of the outside; the mirror cracks, I am an other, and I always was."[31] This reminds us, possibly deliberately, of the monster in the mirror, the theme of Lovecraft's "The Outsider." But this realization of the eeriness of our subjectivity is less shocking for some than it is for others. At the start of the twentieth century W. E. B. Du Bois famously captured the experience of seeing himself through white eyes, seeing himself as a *problem*: "It is a peculiar sensation, this double-consciousness, this sense of always looking at one's self through the eyes of others."[32] Du Bois grew up "away up in the hills of New England, where the dark Housatonic winds between Hoosac and Taghkanic to the sea," perhaps a hundred miles from Providence; S. T. Joshi suggests that Lovecraft's "Miskatonic," the river that runs through Arkham, may well have been inspired by the Housatonic.[33] Du Bois came to realize that "being a problem is a strange experience" in rural Massachusetts. *The Souls of Black Folk* contains weird and eerie metaphors, from the veil that connects and separates the African-American from the white world, to the "second sight" that paradoxically "yields him no true self-consciousness, but only lets him see himself through the revelation of the other world."[34]

Du Bois had already asked Fisher's question: "if we are not who we think we are, what are we?" Similarly Frantz Fanon knew that part of what made him who he was had been provided "by the other, the white man, who had woven me out of a thousand details, anecdotes, stories."[35] A monstrous veil covers these selves. As Black Tom, the antihero of LaValle's novella, says: "Every time I was around them, they acted like I was a monster. So I said goddamnit, I'll be the worst monster you ever saw!"[36]

Is this retreat from the cosmic and the nonhuman no longer weird or eerie? Fisher suggests that agency is eerie when an agent is present that should be absent, or absent when it should be present. The eerie concerns "the forces that govern our lives and the world;" but we should also recognize that "in a globally tele-connected capitalist world ... those forces are not fully available to our sensory apprehension." The impersonal traces of agents distant in space or time possess an eerie quality. One of Fisher's examples of this is capital; "Capital is at every level an eerie entity: conjured out of nothing, capital nevertheless exerts more influence than any allegedly substantial entity."[37] If capital—complex networks of people and things that are experienced as invisible, powerful hyperobjects—can be eerie, then racism might well be too, present in and shaping place, though "not fully available to our sensory apprehension."

A HOSTILE ENVIRONMENT

Perhaps places can be eerily racist despite the absence of a specific agentive source for that racism. In a famous passage in *Black Skin, White Masks* Fanon described a white child on a French train who pointed out Fanon's "frightening" blackness. Fanon thinks, "[a]ll around me the white man... All this whiteness that burns me."[38] Tariq Jazeel notes that Fanon "refers to a whiteness coded through space, in the train carriage, on the platform, at the station, in the city beyond, and so on; the objectification of the black body is inherently spatial."[39] The child and other white passengers produce this effect, but both Fanon and Du Bois were aware of its potential existence in *any* space. As Du Bois explained in 1920:

> I arise at seven. The milkman has neglected me. He pays little attention to colored districts. My white neighbor glares elaborately. I walk softly, lest I disturb him. The children jeer as I pass to work. The women in the street car withdraw their skirts or prefer to stand. The policeman is truculent. The elevator man hates to serve Negroes. My job is insecure because the white union wants it and does not want me. I try to lunch, but no place near will serve me. ... [These things] do happen. Not all each day—surely not. But now and then—now seldom, now, sudden; now after a week, now in a chain of awful minutes; not everywhere, but anywhere—in Boston, in Atlanta. That's the hell of it. Imagine spending your life looking for insults or for hiding places from them...[40]

The potential challenge, refusal, insult, or assault can come at any time, in any place. This can also be felt in the empty landscapes Fisher associated with the eerie. The Black British artist Ingrid Pollard's 1988 work *Pastoral Interlude* consists of a series of five photographic prints, accompanied by brief notes linking the British countryside to racial exclusion and the hidden histories of the slave trade. The caption to an image of Pollard sitting alone on a wall in the Lake District notes, "A visit to the countryside is always accompanied by a feeling of unease, dread."[41] Clearly, racism exists in individual thoughts and acts, but—like Morton's hyperobjects, or any other massively distributed phenomenon—it can be experienced as an impersonal force latent in place even when racist actors are absent.

In this respect Ruff and LaValle turn Lovecraft's weird geography inside out, with African-Americans as the alien outsiders in white America's sundown suburbs and towns. It's hard not to read Lovecraft's "The Shadow over Innsmouth" as a racist description of a town marked by monstrous miscegenation, for example, especially as a character describes the locals' hatred for Innsmouth as "simply race prejudice—and I don't say I'm blaming those that hold it."[42] However, as Ruff says of the description of Lovecraft's white narrator's pursuit by the townsfolk of Innsmouth, "with just a few changes this could easily be the story of a black traveller caught in the wrong place after dark."[43]

VISIBILITY AND MOBILITY

Ballad maps out places of safety and danger. Tommy feels at home in Harlem; "walking through Harlem first thing in the morning was like being a single drop of blood inside an enormous body that was waking up." But as he travels from Harlem to Flushing on the subway, "the further Tommy Tester rode into Queens the more conspicuous he became." The conductor quizzes him about his destination; white passengers note his presence. "He was surveyed but never stopped." When he returns to Flatbush, "the journey felt more threatening because the sun was down," and he is stared at and questioned. Later Tommy thinks about leaving Suydam's house, but he would be like "Satan strolling through Eden" in a white neighborhood at midnight; "He was, in essence, trapped here until morning."[44]

On the other hand, Tommy knows that he can hide his identity from these white observers. As LaValle tells us in the first words of the novella: "People who move to New York always make the same mistake... They come looking for magic, whether evil or good, and nothing will convince them it isn't here. This wasn't all bad, though. Some New Yorkers had learned how to make a living from this error in thinking."[45] Tommy knows this. He can get others—white people—to do what he wants, within limits, by playing roles, being the person they want him to be. "Becoming unremarkable, invisible, compliant—these were useful tricks for a black man in an all-white neighborhood. Survival techniques." Manipulating the veil, "he knew the role bestowed a kind of power upon him. Give people what they expect and you can take from them all that you need."[46] Of course this does not stop him from being watched, harassed, assaulted; but he keeps his head down, drops out of sight, survives. Ruff's characters also exploit white belief in African-American magic. Ruby and Letitia's mothers held séances, faking noises and incidents for white clients. "Afterwards, Momma would laugh and joke about how gullible these people were. White folks' belief that Negroes were magically gifted struck her as the most absurd form of superstition."[47] In both novels magic, like other forms of power, is largely in the hands of white men—but magic is still associated with African-Americans.

Detective Malone tries to control Tommy's movements, telling him "you belong in Harlem, not Red Hook" and threatening him with arrest if he is seen in Brooklyn. Tommy resists; but as Black Tom, in the second half of the novel, he is now fully himself. When Malone meets Tom again, "His demeanor, even his voice, was greatly changed from when they'd last met. The Negro spoke with open disdain and returned Malone's stare so directly that it was Malone who looked away." Tom ignores Malone's order to return to Harlem and replies "You shouldn't be here when I get back." Now it is Malone whose mobility is threatened.[48]

THE JIM CROW MILE

In *Lovecraft Country* travel and safety are key concerns because many of the novel's protagonists are associated with *The Safe Negro Guide* and are attuned to the dangers black travelers face in white America. Horace, the teenage son of the *Guide*'s publisher George Berry, is a talented cartoonist and has marked up the family road atlas, making it "a visual translation of *The Safe Negro Guide*."

> Major Negro population centers like Chicago's South Side were represented as shining fortresses. Smaller neighborhoods and enclaves were marked with towers or oases. Isolated hotels and motels were inns with smiling keepers.

Tourist homes—private residences that lent rooms to Negro travellers—were peasant huts, or tree houses, or hobbit holes."

Less friendly parts of the country were populated by ogres and trolls, vampires and werewolves, wild beasts, ghosts, evil sorcerers, and hooded white knights. In Oklahoma, a great white dragon coiled around Tulsa…

Atticus turned to Massachusetts. Devon County was marked with an icon he'd seen in numerous other places in the atlas: a sundial. Standing beside it, casting his own shadow over the gnomon, was a grim Templar holding a noose."[49]

As the Turners and their friends travel across the US they encounter suspicion, threats, and aggression; they are ordered to move on, warned to get out of town by sundown. Atticus develops a sixth sense. "He knew right away there was going to be trouble in Simmonsville."[50] As Cotton Seiler says: "For black drivers, the road's only constant was uncertainty." Du Bois' description of the "Jim Crow car" made a similar point about rail travel, and Loewen documents the survival of these threats through the twentieth and twenty-first centuries.[51]

Cumulatively, racism deforms space and time, as Ruff notes in the very first words of *Lovecraft Country*, quoting from *The Safe Negro Travel Guide* for 1954:

JIM CROW MILE—A unit of measurement, peculiar to colored motorists, comprising both physical distance and random helpings of fear, paranoia, frustration, and outrage. Its amorphous nature makes exact travel times impossible to calculate, and its violence puts the traveler's good health and sanity constantly at hazard.[52]

Traveling while black is terrifying and exhausting. At one point Montrose bridles at being called an old man, thinking "[a]s for old, well… I'm forty-one. But forty-one, in Jim Crow years, *is* old. Ancient, even." Jim Crow distorts time, taking a toll on black minds and bodies. It adds another kind of disorientation to the distortion of space and time common in fantastic fictions. And it is utterly hostile to black lives. Montrose relives his father's death, Ulysses dying in order to protect Montrose in the 1921 Tulsa lynchings. "He had this look on his face. Horror. Horror at the universe. … That's the horror, the most awful thing: to have a child the world wants to destroy and know that you're helpless to help him."[53]

Ruff's white characters cannot understand this. When he realizes that he has been tricked, Caleb Braithwaite threatens Atticus, saying: "No matter where you go, you'll never be safe." Atticus replies: "What is it you're trying to scare me with? You think I don't know what country I live in? I know. We all do. We always have. You're the one who doesn't understand." This is already the reality of his life: no matter where they go, they will never be safe.[54] Despite this Atticus and his family cast a spell on Caleb Braithwaite that means that he can no longer enter African-American areas, leaving him with a map marked with the black spaces he is now excluded from, a negative print of the *Safe Negro Travel Guide*. Like Black Tom, Atticus has laid claim to space.

CONCLUSIONS: ANSWERING LOVECRAFT

As the above should make clear, both LaValle and Ruff invert Lovecraft's weird geographies in order to challenge his racism. LaValle goes further than this, though, by putting Lovecraft's words (and opinions) into the mouth of Lovecraft's character Robert Suydam. In LaValle's story Suydam states "[a]nd it is my belief that an awful lore is not yet dead," quoting the line from Machen's "The Red Hand" that forms part of the epigraph to "Red Hook." He goes on:

> 'Your people,' Robert Suydam began. 'Your people are forced to live in mazes of hybrid squalor. It's all sound and filth and spiritual putrescence. ... Policemen despair of order or reform, and seek rather to erect barriers protecting the outside world from the contagion.'[55]

This uses phrases from three different parts of Lovecraft's description of Red Hook, adding "*Your people are forced*" to the omniscient narrator's statement that "Red Hook is a maze of hybrid squalor..."[56] Suydam is addressing Tommy—and unlike Lovecraft, he lays the blame for Red Hook's slums elsewhere, by implication at the door of white New York. This critical revision of Lovecraft is met with Tommy's own angry response, as he assumes Suydam is talking about Harlem. Tommy says: "I'm trying to understand what in the hell place you're talking about. It doesn't sound like anywhere I've ever lived."[57] Tommy's response is a straightforward challenge, but it is tempting to read LaValle's revision as a Bakhtinian double-voiced utterance. In one sense all dialogue is double-voiced for Bakhtin, because the speaker's words anticipate the listener's response. However, this example seems closer to what Bakhtin called "internally polemical dis-course—the word with a sideward glance at someone else's hostile word."[58] Suydam, the wealthy white cult leader, has his reasons for sympathizing with the denizens of Red Hook; he tells them "When [the Sleeping King] returns, all the petty human evils, such as the ones visited on your people, will be swept away by his mighty hand."[59] In return for helping him, the ruffians he commands will be saved from this apocalypse. But Suydam's voice still serves a very different aim to that of Lovecraft's original narra-tor, who would blame the area's squalor on the hybridity of its populace. In this way LaValle mimics Lovecraft's text in a spirit of critical parody. Henry Louis Gates argued that the Bakhtinian double-voiced utterance is closely associated with African-American "signifying" and creativity; per-haps LaValle's parody is a stylistic response to Lovecraft's racism.[60]

LaValle also adds another twist to Lovecraft's narrative, addressing the horror of cosmic indifference. Suydam makes a series of promises to the "cutthroats" of Red Hook, which will be realized when the Sleeping King returns. Tommy is not sure this will be much of an improvement.

> The end of this current order, its civilization of subjugation. The end of man and all his follies. Extermination by indifference. ... Maybe yesterday the promise of reward in this new world could have tempted Tommy, but today such a thing seemed worthless. Destroy it all, then hand what was left over to Robert Suydam and these gathered goons? What would they do differently? Mankind didn't make messes; mankind was the mess.[61]

Facing murderous racist violence, hostile policing, and residential discrim-ination, Tommy has a revelation: "A fear of cosmic indifference suddenly seemed comical, or downright naïve... What was indifference compared with malice?" he thinks. "Indifference would be such a relief." Unlike Suy-dam, or Lovecraft, Tommy is not afraid of the indifference of the void, or of the inhuman Sleeping King. He chooses indifference over malice; he tells Malone "*I'll take Cthulhu over you devils any day.*"[62] Lovecraft feared cosmic indifference, because his favorite places still meant something to him; Suydam fears it because he wants there to be something left for him to rule over. Black Tom, on the other hand, has nothing left to lose, and choses alien indifference over human malice.

Alongside Ruff's reminder that there are other ways in which the uni-verse can be horrifying, LaValle's novel suggests that there might be worse things than cosmic indifference—if we consider the different ways in which we can be human alongside the marvels and threats of the nonhuman universe.

NOTES

1. Victor LaValle, *Ballad of Black Tom* (New York: Tor, 2016), 136.
2. Victor H. Green, *The Negro Motorist Green Book* (New York: Victor H. Green & Co., 1949), 1.
3. Matt Ruff, *Lovecraft Country* (New York: Tor, 2016), 13.
4. Carl H. Sederholm and Jeffrey Andrew Weinstock, "Introduction: Lovecraft Rising," in *The Age of Lovecraft*, ed. Carl H. Sederholm and Jeffrey Andrew Weinstock (Minnesota: University of Minnesota Press, 2016).
5. See "Deep Cuts in a Lovecraftian Vein," Bobby Derie, https://deepcuts. blog/.
6. Kij Johnson, *The Dream-Quest of Vellitt Boe* (New York: Tor, 2016). Brian M. Sammons and Oscar Rios, ed., *Heroes of Red Hook* (New York: Golden Goblin Press, 2016).
7. LaValle, *The Ballad*, 130.
8. Cited in Ruff, *Lovecraft Country*, 266.
9. H. P. Lovecraft, "The Crime of the Century," in *Miscellaneous Writings*, ed. S. T. Joshi (Sauk City, WI: Arkham House, 1995), 253, 255; Lovecraft, "In a Major Key," in *Miscellaneous Writings*, 425; and S. T. Joshi, *A Dreamer and Visionary: H. P. Lovecraft in His Time* (Liverpool: Liverpool University Press, 2001), 360.
10. Bennett Lovett-Graff, "Shadows Over Lovecraft: Reactionary Fantasy and Immigrant Eugenics," *Extrapolation* 38, no. 3 (Fall 1997); Silvia Moreno-Garcia, "Magna Mater: Women and Eugenic Thought in the Work of H. P. Lovecraft" (MA dissertation, University of British Columbia, 2016).
11. Lovecraft, "Observations on Several Parts of America," in *Miscellaneous Writings*, 299.
12. Michel Houellebecq, *H. P. Lovecraft: Against the World, Against Life*, trans. Dorna Khazeni (London: Gollancz, 2008), 113.
13. Houellebecq, *H. P. Lovecraft*, 107.
14. Mark Fisher, *The Weird and the Eerie* (London: Repeater, 2016), 17.
15. Carl H. Sederholm and Jeffrey Andrew Weinstock, "Afterword: Interview with China Miéville," in *The Age of Lovecraft*, eds. Carl H. Sederholm and Jeffrey Andrew Weinstock (Minnesota: University of Minnesota Press, 2016).
16. Roger Luckhurst, "The Weird: A Dis/orientation," *Textual Practice* 31, no. 6 (2017).
17. Eugene Thacker, *In the Dust of This Planet: The Horror of Philosophy* (Winchester: Zero Books, 2011); Graham Harman, *Weird Realism: Lovecraft and Philosophy* (Winchester: Zero Books, 2012).

18. Fisher, *The Weird and the Eerie*, 28.
19. James Kneale, "From Beyond: H. P. Lovecraft and the Place of Horror," *Cultural Geographies* 13, no. 1 (January 2006); Farah Mendlesohn, *Rhetorics of Fantasy* (Middleton, CT: Wesleyan University Press, 2008).
20. Thacker, *In the Dust*, 6, 5.
21. Harman, *Weird Realism*, 3.
22. Timothy Morton, *Dark Ecology: For a Logic of Future Coexistence* (New York: Columbia University Press, 2016); Donna Haraway, "Anthropocene, Capitalocene, Plantationocene, Chthulucene: Making Kin," *Environmental Humanities* 6, no. 1 (May 2015), 160.
23. Doreen Massey, "Power-Geometry and a Progressive Sense of Place," in *Mapping the Futures: Local Cultures, Global Change*, eds. John Bird, Barry Curtis, Tim Putnam, and Lisa Tickner (London: Routledge, 1993), 67; Donna Haraway, *Staying with the Trouble: Making Kin in the Chthulucene* (Durham, NC: Duke University Press, 2016), 31.
24. Morton, *Dark Ecology*, 10.
25. Lovecraft, *Selected Letters III, 1929–1931*, eds. August Derleth and Donald Wandrei (Sauk City, WI: Arkham House, 1971), 215. See James Kneale, "'Ghoulish Dialogues': H. P. Lovecraft's Weird Geographies," in *The Age of Lovecraft*, eds. Carl H. Sederholm and Jeffrey Andrew Weinstock (Minnesota: University of Minnesota Press, 2016).
26. Evan Lampe, "In Praise of the Innsmouth Look: Nautical Terror and the Specter of Atlantic History in H. P. Lovecraft's Fiction," *Euramerica* 46, no. 2 (June 2016).
27. David Eltis and David Richardson, *Atlas of the Transatlantic Slave Trade* (New Haven: Yale University Press, 2010).
28. Harman, *Weird Realism*, 125.
29. Fisher, *The Weird and the Eerie*, 61, 11, 11.
30. Lovecraft, *Selected Letters* III, 208.
31. Fisher, *The Weird and the Eerie*, 11–12.
32. W.E.B.Du Bois, *The Souls of Black Folk* (New York: Penguin, 1903), 5.
33. Du Bois, *The Souls*, 4. S.T. Joshi, in Lovecraft, *The Call of Cthulhu and Other Weird Stories* (London: Penguin, 1999), 371.
34. Du Bois, *The Souls, 4,* 5.
35. Frantz Fanon, *Black Skin, White Masks,* trans. Charles Lam Markmann (London: Pluto Press, 1967), 111.
36. LaValle, *The Ballad*, 146–47.
37. Fisher, *The Weird and the Eerie*, 64, 11.
38. Fanon, *Black Skin*, 114.
39. Tariq Jazeel, *Postcolonial Geographies* (London and New York: Routledge, 2019), 129.
40. W.E.B.Du Bois, *Darkwater: Voices From Within The Veil* (New York: Harcourt, Brace And Howe, 1920), 222–23.

41. See http://www.ingridpollard.com/pastoral-interlude.html.
42. H.P. Lovecraft, "The Shadow Over Innsmouth," in *Call of Cthulhu*, 272.
43. Ruff, 'Exploring *Lovecraft Country*', *Lovecraft Country*, 5.
44. LaValle, *The Ballad*, 11–12, 12, 13, 37, 45.
45. LaValle, *The Ballad*, 9.
46. LaValle, *The Ballad*, 12–13, 32.
47. Ruff, *Lovecraft Country*, 220.
48. LaValle, *The Ballad*, 105, 104, 105.
49. Ruff, *Lovecraft Country*, 20.
50. Ruff, *Lovecraft Country*, 31.
51. Cotten Seiler, "'So That We as a Race Might Have Something Authentic to Travel By': African American Automobility and Cold-War Liberalism," *American Quarterly* 58, no. 4 (December 2006), 1099. Du Bois, *Darkwater*, 228–30. James W. Loewen, *Sundown Towns: A Hidden Dimension of American Racism* (New York: The New Press, 2018).
52. Ruff, *Lovecraft Country*, 1.
53. Ruff, *Lovecraft Country*, 273, 293–94.
54. Ruff, *Lovecraft Country*, 365, 366.
55. LaValle, *The Ballad*, 44, 47.
56. H.P. Lovecraft, "The Horror at Red Hook," in *The Dreams in the Witch House and Other Weird Stories* (London: Penguin, 2005), 119–20.
57. LaValle, *The Ballad*, 47.
58. Mikhail M. Bakhtin, *Problems of Dostoevsky's Poetics*, trans. Caryl Emerson (Minneapolis: University of Minnesota Press, 1984), 196.
59. LaValle, *The Ballad*, 57.
60. Henry Louis Gates, *The Signifying Monkey: A Theory of African American Literary Criticism* (Oxford: Oxford University Press, 2014).
61. LaValle, *The Ballad*, 76.
62. LaValle, *The Ballad*, 66, 143; emphasis in the original.

Bibliography

Bakhtin, Mikhail M. *Problems of Dostoevsky's Poetics*. Translated by Caryl Emerson. Minneapolis: University of Minnesota Press, 1984.

Derie, Bobby. "Deep Cuts in a Lovecraftian Vein." Accessed May 14, 2019. https://deepcuts.blog/.

Du Bois, W.E.B. *The Souls of Black Folk*. New York: Penguin, 1903.

———. *Darkwater: Voices From Within The Veil*. New York: Harcourt, Brace and Howe, 1920.

Eltis, David and David Richardson. *Atlas of the Transatlantic Slave Trade*. New Haven: Yale University Press, 2010.

Fanon, Frantz. *Black Skin, White Masks.* Translated by Charles Lam Markmann. London: Pluto Press, 1967.

Fisher, Mark. *The Weird and the Eerie.* London: Repeater, 2016.

Green, Victor H. *The Negro Motorist Green Book.* New York: Victor H. Green & Co., 1949.

Haraway, Donna. "Anthropocene, Capitalocene, Plantationocene, Chthulucene: Making Kin." *Environmental Humanities* 6, no. 1 (May 2015): 159–65.

———. *Staying with the Trouble: Making Kin in the Chthulucene.* Durham, NC: Duke University Press, 2016.

Harman, Graham. *Weird Realism: Lovecraft and Philosophy.* Winchester: Zero Books, 2012.

Houellebecq, Michel. *H.P. Lovecraft: Against the World, Against Life.* Translated by Dorna Khazeni. London: Gollancz, 2008.

Jazeel, Tariq. *Postcolonial Geographies.* London and New York: Routledge, 2019.

Johnson, Kij. *The Dream-Quest of Vellitt Boe.* New York: Tor, 2016.

Joshi, S.T. *A Dreamer and Visionary: H. P. Lovecraft in His Time.* Liverpool: Liverpool University Press, 2001.

Kneale, James. "From Beyond: H. P. Lovecraft and the Place of Horror." *Cultural Geographies* 13, no. 1 (January 2006): 106–26.

———. "'Ghoulish Dialogues': H. P. Lovecraft's Weird Geographies." In *The Age of Lovecraft*, edited by Carl H. Sederholm and Jeffrey Andrew Weinstock, 43–61. Minnesota: University of Minnesota Press, 2016.

Lampe, Evan. "In Praise of the Innsmouth Look: Nautical Terror and the Specter of Atlantic History in H. P. Lovecraft's Fiction." *Euramerica* 46, no. 2 (June 2016): 165–210.

LaValle, Victor. *The Ballad of Black Tom.* New York: Tor, 2016.

Loewen, James W. *Sundown Towns: A Hidden Dimension of American Racism.* New York: The New Press, 2018.

Lovecraft, H.P. "In a Major Key." In *Miscellaneous Writings*, edited by S.T. Joshi, 423–25. Sauk City, WI: Arkham House, 1995.

———. "Observations on Several Parts of America." In *Miscellaneous Writings*, edited by S.T. Joshi, 297–318. Sauk City, WI: Arkham House, 1995.

———. *Selected Letters III, 1929–1931.* Edited by August Derleth and Donald Wandrei. Sauk City, WI: Arkham House, 1971.

———. *The Call of Cthulhu and Other Weird Stories.* Edited by S.T. Joshi. London: Penguin, 2001.

———. "The Crime of the Century." In *Miscellaneous Writings*, edited by S.T. Joshi, 253–55. Sauk City, WI: Arkham House, 1995.

———. "The Horror at Red Hook." In *The Dreams in the Witch House and Other Weird Stories*, edited by S.T. Joshi, 116–37. London: Penguin, 2005.

———. "The Shadow Over Innsmouth." In *Call of Cthulhu and Other Weird Stories*, edited by S.T. Joshi, 268–335. London: Penguin, 1999.

Lovett-Graff, Bennett. "Shadows Over Lovecraft: Reactionary Fantasy and Immigrant Eugenics." *Extrapolation* 38, no. 3 (Fall 1997): 175–92.

Luckhurst, Roger. "The Weird: A Dis/orientation." *Textual Practice* 31, no. 6 (2017): 1041–1061.

Massey, Doreen. "Power-Geometry and a Progressive Sense of Place." In *Mapping the Futures: Local Cultures, Global Change*, edited by John Bird, Barry Curtis, Tim Putnam, and Lisa Tickner, 60–70. London: Routledge, 1993.

Moreno-Garcia, Silvia. "Magna Mater: Women and Eugenic Thought in the Work of H. P. Lovecraft." MA dissertation, University of British Columbia, 2016.

Pollard, Ingrid. "Pastoral Interlude." 1988. http://www.ingridpollard.com/pastoral-interlude.html.

Ruff, Matt. *Lovecraft Country*. New York: Tor, 2016.

Sammons, Brian M. and Oscar Rios, ed. *Heroes of Red Hook*. New York: Golden Goblin Press, 2016.

Sederholm, Carl H. and Jeffrey Andrew Weinstock. "Afterword: Interview with China Mieville." In *The Age of Lovecraft*, edited by Carl H. Sederholm and Jeffrey Andrew Weinstock, 231–43. Minnesota: University of Minnesota Press, 2016.

———. "Introduction: Lovecraft Rising." In *The Age of Lovecraft*, edited by Carl H. Sederholm and Jeffrey Andrew Weinstock, 1–42. Minnesota: University of Minnesota Press, 2016.

Seiler, Cotten. "'So That We as a Race Might Have Something Authentic to Travel By': African American Automobility and Cold-War Liberalism." *American Quarterly* 58, no. 4. (December 2006): 1091–117.

Thacker, Eugene. *In the Dust of This Planet: The Horror of Philosophy*. Winchester: Zero Books, 2011.

CHAPTER 8

The Oceanic Weird, Wet Ontologies and Hydro-Criticism in China Miéville's *The Scar*

Jolene Mathieson

WEIRD WATERS, WET ONTOLOGIES, AND HYDRO-CRITICISM

A complex, four-dimensional space of tempestuous materiality and relentless dynamism often beyond human conceptions of time and scale, the ocean and its masses of churning and drifting waters depreciate and radically deny terrestrial modes of thinking. Recognizing the limits of earlier modes of oceanic thought within the natural and social sciences, the geographers Kimberly Peters and Philip Steinberg have recently called for a way of "thinking *with* the sea that can assist in reconceptualizing our geographical understandings" of the ocean as a "world of flows, connections, liquidities, and becomings" and that can serve as "a means by which the sea's material and phenomenological distinctiveness can facilitate the reimagining and re-enlivening of a world ever on the move."[1] Key to their

J. Mathieson (✉)
University of Hamburg, Hamburg, Germany

© The Author(s) 2019
J. Greve and F. Zappe (eds.), *Spaces and Fictions of the Weird and the Fantastic*, Geocriticism and Spatial Literary Studies,
https://doi.org/10.1007/978-3-030-28116-8_8

wet ontology is a turn away from referencing a stable "plane geometry of points, lines, areas" where matter exists prior to movement[2] and toward adopting a "Lagrangian perspective wherein movement, instead of being subsequent to geography, *is* geography."[3] Lagrangian fluid dynamics is a frame of reference for modeling the movement of water as a fluid parcel traveling through space and time. The motion is determined "by the displacement across space of material characteristics [...that] are known only through their mobility" within the fluid parcel. This means that in opposition to the more common Eulerian perspective wherein movement is modeled by measuring the dynamic forces acting on a stable buoy fixed in space, objects in the Lagrangian perspective "*come into being* as they move (or unfold) through space and time."[4] This also means that space becomes a medium or dynamic force constitutive to the unfolding process.[5]

Peters and Steinberg's foregrounding of Lagrangian fluid dynamics in their wet ontology does not call for a rigid modeling or scientification of oceanic space but rather showcases the ways in which movement has a profound material reality. Their invitation to thinking with the ocean is a transdisciplinary enterprise which also includes the imaginative processes of literature and art. I would thus like to point to the role that *the oceanic weird* plays in facilitating "the reimagining and re-enlivening" of the ocean's ceaseless reconfigurations of material immanence.[6] The oceanic weird—the hodgepodge and mutations of fictions that include maritime science fiction, horror, and fantasy—offers us sophisticated, often self-reflective models of wet ontologies that are invaluable to oceanic studies. These models invite the reader to engage with the chaotic tessellations of maritime assemblages through their representations of water, oceanic space, and marine life, their hybrid and often subversive use of discursive genres, and their interdiscursive translations of knowledge in engineering, seafaring, geography, and the marine sciences. Common to much of the oceanic weird's corpus are varying degrees of soft- and hard-science modes of knowledge production enmeshed with the notions of higher dimensional space as well as the monstrous and the un- and/or "abcanny" (Miéville)—notions that critically serve as salient markers of the ocean's "phenomenological distinctiveness." I argue that the oceanic weird offers fruitful mechanisms for conceptualizing the relations between the knowable and the unknowable in engaging the ocean's material reality.

This reality, Peters and Steinberg show, also entails further critical features. First, the ocean's verticality and material volume as it coalesces "with its movement, its horizontal surface, its angled waves," creates a space "not moved on, but *through* [and] *under.*"[7] Second, the ocean's rhythmic turbulence, resulting from its verticality, material volume, and movement, "gives the sea not only (ever shifting) depth but also *form*," a chaotic tessellation of maritime assemblages "ripe with affective resonances and haptic engagements."[8] Third, the ocean's space "is continually constituted through dynamic re-formations in time, and vice versa," thus enacting a four-dimensional immanence with and through depth, time, and the two dimensions of area.[9]

I would like to add a fourth feature: the hydroelemental force of marine animality. The molecular architecture of water is anomalous compared to other liquids and the reasons for its behavior remains a mystery. But we do know that when the kinetic energy of water molecules is just high enough to disrupt their capacity to form hydrogen bonds and expand (as what happens when water freezes and ice forms), the hydrogen bonds slip and slide over one another, constantly tearing apart and reforming, resulting in liquid water. Water, then, is the emergent property of the picosecond lifetimes of fragile hydrogen bonds engendered *in perpetuum* by the dynamic movement, entropic disorder changes, and structural regularities of water molecules as they interact with one another.[10] While intrinsically unstable, the molecular architecture of water enigmatically introduces enough structure that even at sublimely infinitesimal scales, it is the medium that begets biological matter. In any given one-milliliter-sized drop of seawater, there teems with life upwards of eleven million marine microbes that move, dance, and wiggle, that ooze, leak, and discharge, that reproduce, mutate, get sick, and die.[11] Microbial mélange—viruses, bacteria, protozoans, and algae—substantially enrich and enliven the microscopic movement of watery life.

On a different scale, a growing body of literature in oceanography and marine biology shows that oceanic biomixing—biologically generated turbulence as a result of swimming animals—is both a local phenomenon on part of larger aquatic animals like whales and even jellyfish,[12] as well as a *global* phenomenon on part of tiny organisms, such as zooplankton, that vertically migrate hundreds of meters every day from the ocean's depths to its surface to feed at night. Each individual animal, as they beat their minuscular feathered legs, swim in tandem to form dense aggregations of "tens of meters in vertical extent," potentially creating powerful currents and "considerably altering the physical and biogeochemical structure of the water column."[13] These findings suggest that biological matter, too, is a trans-scalar property fundamentally constitutive to the profound material reality of oceanic movement.

With these four geo-ontological features in tow—verticality and material volume, rhythmic turbulence, four-dimensional immanence, and the hydroelemental force of marine animality—I would like to situate the oceanic weird in an emerging field of hydro-critical scholarship variably known as oceanic studies, blue ecocriticism and blue humanities. This new but already prolific field challenges terracentrism by forging a type of literary and cultural criticism that examines the complex entanglement of cultural mediation, oceanic ecologies and scientific forms of knowledge.[14] This scholarship, however, has yet to study the oceanic weird. I would thus like to exemplarily position China Miéville's oceanic weird novel, *The Scar* (2002), within this field and call attention to the ways oceanic weird geographies and the critical framework of wet ontologies can be mutually enhancing. The oceanic weird is a mode of writing that conceptualizes water, the ocean and the slime materials born from it as mundane matter—but as matter whose wet ontologies are so radical, so alien, that they enact and constitute a *hypermateriality*—a materiality that touches and configures the human but, paradoxically, can seemingly only be accessed speculatively through the metonymic and metaphoric paradigm of the mythological and the unreal.

The Oceanic Weird

Exactly what generic features and mechanisms are at work to make literature "weird," in either the haute or new sense, have been under theorization for at least a decade.[15] In Roger Luckhurst's convincing, and fun, "dis/orientation" of the weird, he identifies the genre's defining feature by its nonexistence as a genre. The weird, he writes, "is not actually there, or only spectrally so. It is better to think of the weird as an inflection or tone, a mode rather than a genre"—or as a verb rather than a noun—that resides and breeds in "the interstices of other forms."[16] "Weirding," then, is a recognition of how literature can "inhere in perversity or transgression," how it "twists or veers away from familiar frames and binary distributions," how it rents apart and unevenly re-stitches the dripping, porous boundaries between genres.[17]

The feeble bonds of the weird and their writhing in but twisting away of traditional tropes have resulted in a catalogue of distorted affects and forms that are recognizable but often evade technical description. The weird is foremost a phenomenological tool for confronting a more-than-human world. It is a way of thinking with but also athwart the reality of a universe deeply inflected by a radical materiality. Originating in the *fin-de-siècle* at a time when scientific discovery began to decenter anthropocentric con-ceptualizes of the universe, the weird's earliest desire, as Luckhurst shows, is its exploitation of "human disgust at formless, structureless, primordial ooze, the slime dynamics that invoke the arche-origins of life itself, a chaos of protozoan mass that dissolves all boundary."[18] The weird is born of "a truly Darwinian traumatism."[19] Its mood is thus thick, palpable, wet. Its atmosphere drips with unease and cognitive distortion.

Although its origins maybe rooted in a certain disgust of the apparent "formlessness" of the primordial, as a (non)genre, it is utterly obsessed with form. The weird is an ungainly hybridization and subversion of generic con-ventions taken from the gothic, fantasy, horror, and science fiction genres, and in many ways defies efforts of categorization and canonization. Inte-gral to these processes, however, is an unconditional avowal of the literary and of the textual. As it corrupts boundaries and rotates in-between genres, the weird also comments on them, self-consciously and meta-discursively articulating a desire to unrationalize narrative through the fuzzy logic of dissonance and rupture. The discursive power of the weird lies in its meta-textuality and its ability to pervert its own conventions.

As part of this perversion, the weird has adopted the monster as its favorite plaything.[20] Taking the uncanny of the gothic, to quote Luckhurst's fanciful phrase, the weird "disorients it in the twist of the abcanny"[21]—the "abcanny" being Miéville's term for the dread born not from a return of the repressed but of the utterly unknowability of the "arche-fossil-as-predator," that presence which touches the human yet moves away, never to be apprehended.[22]

More specific to my delineation of the oceanic weird are three interlock-ing or ambient sets of meaning-making that, I argue, align quite well with both the features listed above as well as wet geographic ontologies. I phrase these sets: "the thinking ocean," "Lucretian slime" and the aforementioned notion of "hypermateriality." With *the thinking ocean*, the oceanic weird activates that long cultural tradition of understanding the ocean, that sub-lime space in the Burkean sense, vast and illimitable, as somehow being alive; that via its currents and tides turning in tandem with the gravita-tional forces of the sun, the moon and the earth, it holds some kind of power over us. This idea is given cultural expression in the mythographic shaping of the ocean as a deity, a divine power often given personified, anthropomorphized features like those of the Ancient Greek Titan god of the sea, Oceanus. In other origin stories and mythologies, the ocean is conceived of metonymically, with zoomorphized forms in the shape of sea creatures residing far down in its shadowy depths, occasionally rising to the surface, such as Ryujin in Japanese mythology.

With *Lucretian slime*, I want to foreground the weird's interest in "slime dynamics" but with reference to Lucretius' ontology of motion. In his *De Rerum Natura*, Lucretius provides a theory of matter that, as Thomas Nail has recently shown, "undermines all the classical ideas of discrete, observable, mechanistic atoms."[23] Like Homer and Thales before him, Lucretius imagines water as the *arche*—the originating or first principle of nature. But instead of metonymically conceptualizing this *arche* as Oceanus, Lucretius uses the figure of Venus as a conduit through which he describes a material ontegenesis of fluidity where the universe's four primary elements—water, earth, fire, and air—are understood as stochastic flows and fluxes that curve and fold and produce emergent bodies. In his account, the turbulent movement of waves disorders laminar flow, gestating bubbles, foam and the primordial ooze of slime—the materialized matter of motion constituted by "haptic field of flows" as they fold or curve in on themselves.[24]

And with *hypermateriality*, I want to "disorientate" Timothy Morton's theory of the "hyperobject." In *Hyperobjects: Philosophy and Ecology at the End of the World* (2013), Morton's theorizes hyperobjects as entities massively distributed in space and time that disturb the human but cannot be fully apprehended. Whether objects can remain discreet as singular entities at this scale and dimension is questionable. And because objects, when thinking with the ocean, come into being as they unfold in the space and time of movement, I mutate this notion into one of hypermateriality. Hypermaterials, like hyperobjects, are viscous—just like the properties of H_2O, they are sticky and adhere to all substances and entities they touch, no matter the degree of resistance.[25] They are phased—hypermaterials are situated in but also encompass higher dimensional space, so that they require our focused attention to come into view. And, they are inter-material—complex, dynamic formations forged by the interrelations of more than one material.[26]

Because of the ontological ability of hypermaterial to exist longer than human modes of material existence and because hypermaterial can destroy matter at catastrophic scales, it cannot be entirely apprehended, thus readily lending itself to be conceptualized in either nonmaterial terms or in hybrid terms of monstrous materiality. And this is the paradox: the hypermaterial is so radically real that it can't be fully comprehended by the empirical models available to us. Instead, our inhered epistemic response is to conceive of the reality of the hypermaterial through the metonymic and metaphoric paradigm of the unreal, the supernatural, and the abcanny. This is why the very real dangers at sea have been allegorically personified in the figure of the pychopompos siren, and why the plant and animal life-forms residing in the depths of the ocean, often drudged up to the surface by the trawling nets of fisherman or washed ashore in a bilious, bloated mass of blubber and mucus, have long fed into the cultural imagination as malevolent sea monsters.

Oceanic Thinking in China Miéville's *The Scar*

A particularly rich and complex model of oceanic weird wet ontology is Miéville's *The Scar*, an 800-page steampunk-fantasy novel set in the author's Bas-Lag universe. A remarkably dense and fully realized geography of landmasses, littoral zones, and seas, full of aquatic races and monsters, of pirates, and floating ship cities, the novel's complex narrative structure begins with a five-page set piece that builds a world constructed by a deep knowledge and the interdiscursive translations of contemporary oceanography and marine biology. From the outset, the oceanic world forged in *The Scar* is in many respects recognizably our own. "The sea throngs"—it is a space of relentless emergent dynamism and flux, fluidity, and non-locality:

> It has been given many names. Each inlet and bay and stream has been classified as if it were discreet. But it is one thing, where borders are absurd. It fills the spaces between stones, curling around coastlines and filling trenches between continents.[27]

The sea is the connective tissue that cradles the terrestrial. It is the medium whose movement shapes geography and forges space. It is the materialized matter of motion, of foam and primordial materials that throw out "rude creatures" which "emit slime and phosphorescence and move with flickering of unclear limbs." The hypermateriality of the ocean and the alterity of the creatures in its depths may invite us to monster them, as the "logic of their forms derives from nightmares."[28] But when frozen at "the edges of" Bas-Lag, the sea simulates the terrestrial, confusing and complicating its supposed alterity: "Huge slabs of frozen sea mimic the land, and break and crash and reform, criss-crossed with tunnels, the homes of frost-crabs, philosophers with shells of living ice."[29] The sea, shifting and curving, moves from liquid to solid states as H_2O molecules move closer together, holding and hugging the ocean's volume into tighter forms, into living platforms of philosophy. The sea, in this sense, is a "theory machine," an entity that invites theoretical formulation.[30] As the oceanic anthropologist Stefan Helmreich has wisely warned us, using the ocean as a theory machine can easily undermine the conceptual purpose of thinking through and with the ocean as a living entity beyond human epistemes.[31] But a robust wet ontology, one in which literature and cultural mediation also play a role, must be sensitive to the ways in which the ocean's materiality has been discursively inscribed, in and under, by texts, and how this in turn informs terrestrial ontologies.

The novel's first chapter opens with Bellis Coldwine, a linguist and author, as she sails away in exile from New Crobuzon, her home city grown rich through trade and slavery, to one of its colonies across the Swollen Ocean. Intertextually echoing the narrative tropes of sea adventure novels like Edgar Allan Poe's *The Narrative Arthur Gordon Pym of Nantucket*, which famously incorporated passages from a number of expedition narratives such as Jeremiah N. Reynolds' *Address on the Subject of a Surveying and Exploring Expedition to the Pacific Ocean and the South Seas*,[32] the novel's heterodiegetic narrative mode is interspersed with epistolary sections in which Bellis catalogues her anxieties as well as scientific interests while "staring at the endless green waves with distaste."[33]

Bellis, in many ways, represents the clumsy anthropocentric attitudes of twentieth-century thinkers toward the ocean. W. H. Auden, for example, denies the sea inherent form and meaning, seeing it as the mere "symbol for the primordial undifferentiated flux, the substance which became created nature only by having form imposed upon or wedded to it."[34] Carl Schmitt, in a more crude rejection, describes it as a space that "has no *character*, in the original sense of the word, which comes from the Greek *charassein*, meaning to engrave, to scratch, to imprint. On the waves there is nothing but waves."[35] Bellis, too, mostly projects her terrestrial alterity onto the ocean's apparent horizontality: "The open sea. Waves like insects in incessant motion. Stunning and empty."[36] And this, in spite of having previously visited Salkrikaltor City, a sprawling underwater metropolis beyond "unseen rock and sand," inhabited by the Cray, an aquatic race with humanoid upper bodies, protruding gills behind the ears, and rock lobster lower bodies.[37] Salkrikaltor City is a fully realized aqueous city whose medium, the currents and waves, gives it form, where "every motion seemed stretched out in the water," where "Cray swam over roofs, flapping their tails in an inelegant motion," where "currents picked at organic rubbish mouldering in coral courtyards."[38] Despite her dismissal of the ocean, upon observing "the curves of great predators like orca," Bellis becomes "briefly fascinated with submarine carnivores." The naturalist on board, Johannes Tearfly, is all too happy to rummage through his books on "*Sardula Anatomy; Predation in Iron Bay Rockpools; Theories of Megafauna*" and show her descriptions of scientific discoveries and "sensational depictions of ancient, blunt-heeded fish thirty feet long, of global sharks with ragged teeth and jutting foreheads."[39]

Bellis' fascination with monstrous sea creatures activates a long literary tradition of weirding the ocean that becomes most manifest in antecedent textual representations of the kraken—one of the most productive artifacts of scientific sensationalism and favorite monster of the weird. A fictional version of the giant squid, the kraken was famously monstered in Jules Verne's *Twenty Thousand Leagues Under the Sea* in a scene that sees Captain Nemo and his crew on the Nautilus attacked by a group of highly intelligent giant squid, "horrible monsters, worthy of appearing in any teratological legend," who feel threatened by the invasion of the submarine within their waters.[40] Verne's kraken scene is written in short, parataxical sentences that zoologically describe the squid as scientific specimen but semanticize them through the use of exclamation marks, question marks, and the enumeration of vast measurements as the unknowable, othered 'it' with alien qualities—"what a freak of nature: a bird's beak on a mollusc!"[41] Defying taxonomy, the kraken is a monster whose material transgression against boundaries and gelatinous, massively mucosal tentacles horrify the crew. As an example of scientific sensationalism, the kraken passage in this novel is important in so far that it serves as the template upon which oceanic weird imaginings have been grafted. Its "limb-tongue," as Luckhurst thinks of it, suggests utter alterity, which is why weird writers have adopted the tentacle as "the emblem of that which will not correlate," that which cannot "be reduced to categories of human thought."[42]

In counterargument to the supposed absolute alterity of oceanic hypermateriality, however, is John Wyndham's *The Kraken Wakes* (1953), which depicts a war between amorphous, cephalopod-like creatures residing in the depths of the ocean's cavernous interior and the British government, who, upon discovery of the kraken, insist on disrupting their living space. By the end of the novel, the polar ice caps have melted, the world has flooded, irreversible catastrophic climate change has fundamentally changed the face of the Earth, and the human population has been decimated. This novel warns us about our post-Darwinian tendencies to inscribe the aesthetics of interspecies othering through the formal mode of scientific sensationalism, a deceptive textual strategy that "requires knowledge, research, careful timing, and, if possible, some literary ability" in order to "avoid the suggestion of sensationalism for sensationalism's sake."[43] By sensationalizing and monstering natural-occurring phenomena, we emote them, we derange, we charge them with an exciting, palpable, but addictive energy. Sensationalism, like a living organism but as a discursive practice based on inheritable *topoi*, easily gives rise to memes and mutations and variation upon variation in which, in this particular example, the sentient kraken becomes ever more monstered and alienated. So much so, that Miéville, in another novel, aptly titled *Kraken: An Anatomy* (2010), parodies the campy, preposterous but commonplace potpourri of the mystical and the scientific in science fiction and weird practice. Here, the famed specimen of giant squid located in London's Natural History Museum Darwin Centre magically disappears, is worshipped by an archaic sect of squid worshippers called the Congregation of God Kraken, is animated from its 10% formol-saline solution to full seawater embryonic God, and, upon animation, initiates a deluge apocalypse upon which the world becomes inundated in water, subjugating the terrestrial to an unmitigated wet ontology. As this eschatological weirded parody attests, narrativizing nature and the re-narrativization of our confused emotive responses to oceanic ontologies can alienate nature to absurd degrees. And it may one day take its revenge.

In the act of representing and sensationalizing our fear or our misapprehension of nature, we forget, for example, that the kraken, although sentient and agentic and somehow uncanny with its profound, intelligent eyes, is also vulnerable to changes in its environment, to elemental forces that act upon it and destroy it. Both *The Kraken Wakes* and *Kraken: An Anatomy* function as parables, as warnings of our discursive forming power in which we monster animals, especially of those thriving in environments inhospitable and thus "alien" for *Homo sapiens*. By alienating such natural environments and the species that live in them, as residents of a shared ecosystem, we are, in fact, monstering and alienating ourselves. And as science fiction/weird novels based on the epistemic tradition of monstering that which cannot be easily accessed, they also implicate themselves in this process.

The weird though, as a genre that corrupts its own conventions, also weirds itself. And so, while the monster may be abcanny, it can be twisted away from the unknowable *into* the understandable, maybe even the lovable, and as *The Scar* argues, the monster can certainly enhance wet ontology. Also on board with Bellis and Dr. Tearfly, are the so-called Remade, prisoners of New Crobuzon's authoritarian government which have been punished through monstering—savagely transformed into hybrid entities from the animal, vegetable and mechanical kingdoms—and then condemned to social death as slaves. Tanner Sack, one such Remade, is punished by having the "alien tissue" *par excellence*—large tentacles—grafted to his chest that "itched and shed skin," that lay limp, awkward and undead, rendering him useless and othered from himself.[44] When pirates commandeer the ship for Armada—a floating pirate city thousands of years old, welded from hundreds of vessels and the odds and ends of material and ideological detritus, which have been absorbed like "mindless plankton"—the Remade are granted political life.[45] Performing work that "he had always done—patching and repairing, rebuilding, fumbling with tools by great engines," Tanner is now employed as a deep-sea engineer, "in the crush of water, watched by fishes and eels, buffeted by currents born miles away."[46] His tentacles, which earlier "had hung like stinking dead limbs" begin to move and "new sensations passed through them, and their sucker pads flexed gently and attached themselves on surfaces nearby" and he learns to "move them by choice."[47] Water and its Lucretian slime dynamics, encasing and curling around Tanner, becomes the recuperative medium that bestows him *bios*—full life. Fully at home in a materiality one conceived as alien, his "sense that things were wrong, dangerous or uncertain dissipated with currents."[48] Tanner celebrates his newfound *bios* by undergoing a further set of body modifications so that he can become fully amphibious. By reappropriating monstrosity, his bubbling, oozing hybridity marks the decussation of the ocean's materiality with global political economies and territories. And by thinking through and with the ocean, he shows us that we can weird the weird, or work *athwart theory*, as Helmreich recommends, by twisting in and out of the heuristic and phenomenological mechanisms of theory machines.[49] The oceanic weird can curve and unfold texts to discover monsters that are not mere alterity—they are the hopeful monsters of the hypermaterial.

Another breed of hopeful monsters, but much more horrifying, are the deep-sea dwelling grindylow, humanoid-viperfish assemblages that notoriously resist New Crobuzon's colonization efforts in the depths of the Cold Claw Sea. Hunting down the Armada to find a spy who infiltrated their home waters and created maps outlining their territory, the grindylow sift through the oceans' currents asking the sea to help find the floating city that is "leaving lies and stories," cruel little theory machines of terrestrial geography threatening their wet ontology, "eddying behind it."[50] They follow "little filigrees of rumor" that "twist away from them," that "recurve and tease."[51] They encounter "monkish intelligences masquerading as cod and congers," "salinae, the brine elementals" that communicate with "liquid shrieks," and whales, who they harness and ride.[52] Upon making contact with the Armada, they cruelly tear away at the skin of its bio-engine, an avanc, a gentle godwhale who has propelled the Armada into the dangerous waters of the Hidden Ocean. They inflict injuries "like an excavation: a wide, ragged trench thirty feet deep and many yards long," its inner walls "a crumbling mess of shattered cells, fouled with the residue" of "oily pus."[53] As Dr. Tearfly and his team of scientists inspect the avanc in hope of healing it, they are dumbstruck by its scale, "humbled by the geography of their patient" and horrified to see "clots of the semi-liquid" tissue twist away from it and rise, "strings of matter stretching and snapping behind them."[54] A broken mass of oozing, bloody purulence, the ocean's tender godwhale materially embodies the consequences of a terrestrial attitude that refuses to think with and through the ocean. Despite their savagery, the grindylow are both horrifying and hopeful in the sense that they do not represent utter material alterity but operate athwart terrestrial logics of colonization.

The Hidden Ocean is a place full of boiltides and sentient whirlpools, and home to The Scar—the "face of water, colors, and eddies moving in strata," "extending down *miles*" that gives the novel its name.[55] The allegorical strategies of the novel and its weirding of them, renders The Scar many things. It is at once the hypermaterial space of higher dimensions, folded and unfolded spacetime, that the leaders of Armada want to breach in order to amass new powers. It is exemplified wet ontology—verticality and material volume, rhythmic turbulence, four-dimensional immanence, and hydroelemental animality—a material reality that tears and leaks through the terrestrial. But it is also the legacy of literary and scientific discursivity: the so-called "scars" used in geological nomenclature to designate the deformations and indentions of the earth's mantle, and the "scarred ocean floor" of Peter Watts' weird hard-science fiction novel *Starfish*, the rut of deep hydrothermal vents and volcanoes traversing the planet, where water molecules are heated to originate life.[56] At the end of *The Scar*, Bellis learns that, primarily, "scars are memories,"[57]—that they are the material manifestations of a planetary ontology constituted through movement which must endure the wounds of collision.

While the thinking ocean, Lucretian slime, and hypermateriality all take recourse to the divine, the supernatural or the monstrous in order to structure and relay meaning, they are nonetheless the result of anthropomorphic, acculturated *topoi* with a long diachronic reach. This novel weirds or works athwart their discursive shaping power: it both uses these modes of thought as cognitive tools inhered in human culture meant to help us confront and contextualize a world that moves beyond the limitations of terrestrial and representationalist thought, but also twists them away from their familiar frames to un-other the monster and remind us of their danger. The complexity and integrity of *The Scar*'s weird geographies, while only briefly touched upon here, open to us an ingress for a wet ontology. And it further shows us how the modalities of the oceanic weird—the thinking ocean, Lucretian slime, and hypermateriality—can be implemented as dynamic modes for thinking *with* the ocean, a space of materiality that is very much with us in this world.

NOTES

1. Philip Steinberg and Kimberly Peters,"Wet Ontologies, Fluid Spaces: Giving Depth to Volume Through Oceanic Thinking," *Environment and Planning D: Society and Space* 33 (2015): 248; emphasis in the original.
2. Ibid. See also Doreen Massey, *For Space* (Thousand Oaks, CA: Sage, 2004), whose work Peters and Steinberg heavily draw on.
3. Steinberg, "Of Other Seas: Metaphors and Materialities in Maritime Regions," *Atlantic Studies* 10 (2013): 160; emphasis in the original.
4. Ibid., 161; emphasis added.
5. Ibid.
6. Steinberg and Peters, "Wet Ontologies," 248.
7. Steinberg and Peters,"Wet Ontologies," 253; emphasis in the original.
8. Ibid., 250; emphasis in the original.
9. See Steinberg, "Of Other Seas," 160. In Steinberg and Peters, "Wet Ontologies," published at a later date, he/they drop the four-dimensionality of oceanic space without explanation, but because of its resonances with the higher dimensionality of the hypermaterial, I have chosen to keep it. See the section "The Oceanic Weird" of this article.
10. See Philip Ball, "Water—An Enduring Mystery," *Nature* 452 (March 2008): 291–92; as well as Ursula Goodenough and Terrence W. Deacon, "The

Sacred Emergence of Nature," in *The Oxford Handbook of Religion and Science*, ed. Philip Clayton (Oxford: Oxford University Press, 2008), 855–56.

11. Carlo Heip, et al., *Marine Biodiversity and Ecosystem Functioning* (Dublin: MarBEF, 2009), 19.

12. Trish Lavery, et al., "Can Whales Mix the Ocean?," *Biogeosciences Discussions* 9 (2012): 8387–403; Kakani Katija and John Dabiri, "A Viscosity-Enhanced Mechanism for Biogenic Ocean Mixing," *Nature* 460 (July 2009): 624–26.

13. Isabel A. Houghton, et al., "Vertically Migrating Swimmers Generate Aggregation-Scale Eddies in a Stratified Column," *Nature* 556 (2018): 497.

14. See, for example, Helen M. Rozwadowsky, *Fathoming the Ocean: The Discovery and Exploration of the Deep Sea* (Cambridge, MA: Belknap, 2005); Margaret Cohen, *The Novel and the Sea* (Princeton University Press, 2012); Hester Blum, "Introduction: Oceanic Studies," *Atlantic Studies* 10, no. 2 (2013): 151–55; Ursula Kluwick and Virginia Richter, eds., *The Beach in Anglophone Literatures and Cultures: Reading Littoral Space* (Farnham: Ashgate, 2015); Elizabeth DeLoughrey, "Submarine Futures of the Anthropocene," *Comparative Literature* 69, no. 1 (2017): 32–44; Stacy Alaimo, "Violet Black," in *Prismatic Ecology: Ecotheory Beyond Green*, ed. Jeffrey Jerome Cohen (Minneapolis: University of Minnesota Press, 2013), 233–51; and Steve Mentz and Martha Elena Rojas, eds., *The Sea and Nineteenth-Century Anglophone Literary Culture* (London: Routledge, 2017).

15. See, for example, Benjamin Noys and Timothy S. Murphy, "Introduction: Old and New Weird," *Genre* 49, no. 2 (2016): 117–34; Roger Luckhurst, "In the Zone: Topologies of Genre Weirdness," in *Gothic Science Fiction 1980–2010*, eds. Sara Wasson and Emily Alder (Liverpool: Liverpool University Press, 2011), 21–35; and China Miéville, "M.R. James and the Quantum Vampire: Weird; Hauntological: Versus and/or and and/or or?," *Collapse IV* (2008), 105–28.

16. Roger Luckhurst, "The Weird: A Dis/Orientation," *Textual Practice* 31, no. 6 (2017): 1045.

17. Ibid., 1052.

18. Ibid., 1054.

19. Ibid.

20. On monstrosity in the haute weird, see Emily Alder, "(Re)encountering Monsters: Animals in Early-Twentieth-Century Weird Fiction," *Textual Practice* 31, no. 6 (2017): 1083–1100.

21. Luckhurst, "Dis/Orientation," 1053, 1054.

22. Miéville, "M.R. James and the Quantum Vampire", 113.

23. Thomas Nail, *Lucretius I: An Ontology of Motion* (Edinburgh: Edinburg University Press, 2018), 21.

24. See ibid., 31.

25. Timothy Morton, *Hyperobjects: Philosophy and Ecology at the End of the World* (University of Minnesota Press, 2013), 1.
26. Ibid.
27. Miéville, *The Scar*, 1.
28. Ibid.
29. Ibid.
30. See Stefan Helmreich, *Sounding the Limits of Life: Essays in the Anthropology of Biology and Beyond* (Princeton, NJ: Princeton University Press, 2016), 95.
31. Ibid, 96–99.
32. Edgar Allan Poe, *The Narrative Arthur Gordon Pym of Nantucket* [1838], ed. J. Gerald Kennedy (Oxford: Oxford University Press, 2008), 120–25.
33. Miéville, *Scar*, 69.
34. W.H. Auden, *The Enchafèd Flood, or, The Romantic Iconography of the Sea* [1951] (London: Faber & Faber, 1985), 16–17.
35. Quoted in Steinberg and Peters, "Wet Ontologies," 248–49; emphasis in the original. See Carl Schmitt, *The Nomos of the Earth in the International Law of the Jus Publicum Europaeum* [1950], trans. G.L. Ulmen (New York: Telos, 2003), 42–43.
36. Miéville, *Scar*, 103.
37. Ibid., 51.
38. Ibid., 52.
39. Ibid., 41.
40. Jules Verne, *Twenty Thousand Leagues Under the Sea* [1870], trans. William Butcher (Oxford: Oxford University Press, 2009), 344.
41. Ibid.
42. Luckhurst, "Dis/Orientation," 1054. See also Eugene Thacker, *Tentacles Longer Than Night: Horror of Philosophy, Vol. 3* (Winchester: Zero Books, 2015).
43. John Wyndham, *The Kraken Wakes* (London: Michael Joseph, 1953), 234.
44. Miéville, *Scar*, 37.
45. Ibid., 207.
46. Ibid., 117.
47. Ibid., 116.
48. Ibid., 662.
49. Helmreich, *Sounding the Limits*, 104.
50. Miéville, *Scar*, 248.
51. Ibid.
52. Ibid., 249.
53. Ibid., 685.
54. Ibid., 685–86.
55. Ibid., 753; emphasis in the original.
56. Peter Watts, *Starfish* (New York: Tor, 1999), 302.
57. Miéville, *Scar*, 791.

Bibliography

Alaimo, Stacy. "Violet Black." In *Prismatic Ecology: Ecotheory Beyond Green*, edited by Jeffrey Jerome Cohen, 233–51. Minneapolis: University of Minnesota Press, 2013.

Alder, Emily. "(Re)encountering Monsters: Animals in Early-Twentieth-Century Weird Fiction." *Textual Practice* 31, no. 6 (2017): 1083–100.

Auden, W. H. *The Enchafèd Flood, or, The Romantic Iconography of the Sea*. 1951. London: Faber & Faber, 1985.

Ball, Philip. "Water—An Enduring Mystery," *Nature* 452 (March 2008): 291–92.

Blum, Hester. "Introduction: Oceanic Studies." *Atlantic Studies* 10, no. 2 (2013): 151–55.

Cohen, Margaret. *The Novel and the Sea*. Princeton University Press, 2012.

DeLoughrey, Elizabeth. "Submarine Futures of the Anthropocene." *Comparative Literature* 69, no. 1 (2017): 32–44.

Goodenough, Ursula, and Terrence W. Deacon. "The Sacred Emergence of Nature." In *The Oxford Handbook of Religion and Science*, edited by Philip Clayton, 853–71. Oxford: Oxford University Press, 2008.

Heip, Carlo, et al. *Marine Biodiversity and Ecosystem Functioning*. Dublin: MarBEF, 2009.

Helmreich, Stefan. *Sounding the Limits of Life: Essays in the Anthropology of Biology and Beyond*. Princeton, NJ: Princeton University Press, 2016.

Houghton, Isabel A., et al. "Vertically Migrating Swimmers Generate Aggregation-Scale Eddies in a Stratified Column." *Nature* 556 (2018): 497–500.

Katija, Kakani, and John Dabiri. "A Viscosity-Enhanced Mechanism for Biogenic Ocean Mixing." *Nature* 460 (July 2009): 624–26.

Kluwick, Ursula, and Virginia Richter, eds. *The Beach in Anglophone Literatures and Cultures: Reading Littoral Space*. Farnham: Ashgate, 2015.

Lavery, Trish, et al. "Can Whales Mix the Ocean?" *Biogeosciences Discussions* 9 (2012): 8387–403.

Luckhurst, Roger. "In the Zone: Topologies of Genre Weirdness." In *Gothic Science Fiction 1980–2010*, edited by Sara Wasson and Emily Alder, 21–35. Liverpool: Liverpool University Press, 2011.

———. "The Weird: A Dis/Orientation." *Textual Practice* 31, no. 6 (2017): 1041–61.

Massey, Doreen. *For Space*. Thousand Oaks, CA: Sage, 2004.

Mentz, Steve, and Martha Elena Rojas, eds. *The Sea and Nineteenth-Century Anglophone Literary Culture*. London: Routledge, 2017.

Miéville, China. "M.R. James and the Quantum Vampire: Weird; Hauntological: Versus and/or and and/or or?" *Collapse IV* (2008): 105–28.

———. *The Scar*. 2002. London: Macmillan, 2011.

———. *Kraken: An Anatomy*. London: Macmillan, 2010.

Nail, Thomas. *Lucretius I: An Ontology of Motion*. Edinburgh: Edinburgh University Press, 2018.

Noys, Benjamin, and Timothy S. Murphy. "Introduction: Old and New Weird." *Genre* 49, no. 2 (2016): 117–34.

Poe, Edgar Allan. *The Narrative Arthur Gordon Pym of Nantucket*. 1888. Edited by J. Gerald Kennedy. Oxford: Oxford University Press, 2008.

Rozwadowsky, Helen M. *Fathoming the Ocean: The Discovery and Exploration of the Deep Sea*. Cambridge, MA: Belknap, 2005.

Schmitt, Carl. *The Nomos of the Earth in the International Law of the Jus Publicum Europaeum*. 1950. Translated by G. L. Ulmen. New York: Telos, 2003.

Steinberg, Philip, and Kimberly Peters. "Wet Ontologies, Fluid Spaces: Giving Depth to Volume Through Oceanic Thinking." *Environment and Planning D: Society and Space* 33 (2015): 247–64.

Steinberg, Philip E. "Of Other Seas: Metaphors and Materialities in Maritime Regions." *Atlantic Studies* 10 (2013): 156–69.

Thacker, Eugene. *Tentacles Longer Than Night: Horror of Philosophy, Vol. 3*. Winchester: Zero Books, 2015.

Verne, Jules. *Twenty Thousand Leagues Under the Sea*. 1870. Translated by William Butcher. Oxford: Oxford University Press, 2009.

Watts, Peter. *Starfish*. New York: Tor, 1999.

Wyndham, John. *The Kraken Wakes*. London: Michael Joseph, 1953.

CHAPTER 9

"Through the Eyes of Area X": (Dis)Locating Ecological Hope via New Weird Spatiality

Gry Ulstein

Mark Fisher has noted that every account of the weird must start with H. P. Lovecraft.[1] This chapter keeps Lovecraft as an important comparative historical and thematic coordinate, but starts where Lovecraftian weirdness usually ends: in nihilism and existential dread. Famously, Lovecraft's weird tales thematize the maddening insignificance of the human individual in the encounter with the unknown. Considering this theme's revival in the twenty-first-century crisis-ridden, apocalyptic mentality, several contemporary scholars have pointed out the critical potential of Lovecraft's work for ontologies such as posthumanism and speculative realism.[2] This development is part of a broader ontological shift concerned with decentering the human. The recent "nonhuman turn" formulated by Richard Grusin in 2015 involves critical voices within ecological thought such as Stacey Alaimo, Jane Bennett, Rosi Braidotti, Donna Haraway, Bruno Latour, and Timothy Morton. They call for a radical change in the way humans imagine

G. Ulstein (✉)
Ghent University, Ghent, Belgium

© The Author(s) 2019
J. Greve and F. Zappe (eds.), *Spaces and Fictions of the Weird and the Fantastic*, Geocriticism and Spatial Literary Studies, https://doi.org/10.1007/978-3-030-28116-8_9

and negotiate their position in the world—a call made more urgent in the context of Anthropocene issues like global warming.[3]

Brad Tabas suggests that the "horror of the Anthropocene and the real of weird realism remind us of the inutility of trying to naively suture together the real and the Natural, even if this is clearly the ambition of most place-based writing and critical studies thereof."[4] Although "weird realism" might sound like a contradiction in terms, Fisher and Graham Harman agree with Tabas that the weird's intentionally exaggerated focus on the real—in contrast to traditional fantasy—serves to heighten the "irruption into *this* world of something from the outside."[5] As James Kneale also notes, the weird encourages "the reader to see *reality itself* as weird."[6] Emanating from this peculiar form of realism, Lovecraft's cosmic dread "silence[s] the dominant human,"[7] and can make sobering interjections into anthropocentrism and individualism. But the traditional weird leans heavily on nihilism, and seems uninterested in whether the human silencing might allow other realities to resonate. It does not typically evoke affects like hope or affirmation, which Alexa Weik von Mossner suggests might be crucial "to even the most gruesome views of the future"—or the present.[8] Benjamin Noys and Timothy S. Murphy argue that the more recent wave of weird fiction called "New Weird" conversely displays "a new sensibility of welcoming the alien and the monstrous as sites of affirmation and becoming."[9]

While maintaining a focus on paradoxical spaces and cosmic dread also found in the traditional weird, New Weird writers such as Jeff Vander-Meer and China Miéville more explicitly explore the extent to which the weird's ruptured spaces can offer useful insights into nonhuman realities. I therefore situate my argument alongside, on the one hand, Erin James' econarratology, which seeks to complicate "our definition of 'environment' beyond familiar tropes such as wilderness or the pastoral."[10] On the other hand, I draw on Timothy Morton's "dark ecology," which invites exploration of "all kinds of art forms as ecological: not just the ones that are about lions and mountains, not just journal writing and sublimity. The ecological thought includes negativity and irony, ugliness and horror."[11] Accordingly, my point of departure in this chapter is that recent weird narratives revisit Old Weird intrusions of monstrosity into normality and, via various conventional and unconventional narrative techniques, ask questions about the aftermath: what comes *after* the weirding of reality, besides existential dread?

Lovecraft's oft-quoted claim that the "oldest and strongest emotion of mankind is fear, and the oldest and strongest kind of fear is fear of the unknown,"[12] reflects the core of the Old Weird and its fearful fascination with the unknown outside. According to Rebecca Solnit, however, hope is also intimately bound up with the unknown:

> Hope locates itself in the premises that we don't know what will happen and that in the spaciousness of uncertainty is room to act. When you recognize uncertainty, you recognize that you may be able to influence the outcomes— you alone or you in concert with a few dozen or several million others. Hope is an embrace of the unknown and the unknowable.[13]

Hope as an embrace of the unknown and unknowable reads at once as a useful interjection into the Anthropocene discourse and into the cosmic horror of weird fiction. The unknown for Lovecraft equals unplumbed space: eternal and exhaustive beyond human comprehension, and consequently beyond human influence. The analogies in old-style weird narrative seem to offer little, therefore, besides fear and a humbling of the human condition. New Weird stories, on the other hand, often retain a sense that it is possible to navigate the unknown horrors; they entertain a strange form of hope for an existence in a world even after—or even through—cosmic dread. Using Jeff VanderMeer's critically acclaimed Southern Reach trilogy as a case study, this chapter explores how the weird, via this peculiar hope, can productively question spatial categories like environment, nature, and wilderness, and thereby challenge readers to resituate their "normal" way of thinking in the increasingly weirded time of the Anthropocene.

Weird Thresholds

Lovecraft's weird i s acutely spatially oriented, with unfathomable cosmos and non-Euclidian geometry encroaching violently upon familiar places such as the New England countryside, which often provides the settings for these stories. This spatial focus is key to what Lovecraft sees as the most important component of the weird tale in order to achieve a sense of cosmic horror: an "essence of real externality."[14] Lovecraft's mythos consists of creatures too terrible to be contained within the human mind— awe-inspiring and dreadful not only because they appear as monsters with supernatural powers, but because they are too massively *other*: they twist the human perception of the world to such an extent that the very laws of nature break down. In order to engage with the weird, therefore, Lovecraft claims "we must remember to leave our humanity and terrestrialism at the threshold."[15] Lovecraft's use of the spatial metaphor of a threshold is interesting in itself. It gives the sense that writing and reading weird tales involve conscious movement across an exposed border between two spaces: the terrestrial space and the nonhuman/cosmic space. Fisher argues that this threshold suggests a "contained or localized realism" that is necessary for Lovecraft's weird in particular: "By setting his stories in New England rather than in some inviolate, far-distant realm [as opposed to fantasy], Lovecraft is able to tangle the hierarchical relationship between fiction and reality."[16]

The representation of the environment in weird fiction is intensely spatial, but also counterintuitive in the sense that it does not behave according to human expectations of space. "The most radical thematizations of space," Marie-Laure Ryan notes, "are those that involve alternative or logically inconsistent worlds. [...] Logically impossible story spaces are the narrative equivalent of M.C. Escher's pictorial representations of worlds that violate the laws of perspective."[17] In terms of form, the weird therefore corresponds somewhat with the concept of "unnatural narrative," defined by Brian Richardson as "those texts that violate mimetic conventions by providing wildly improbable or strikingly impossible events; they are narratives that are not simply nonrealistic but antirealistic."[18] Although Richardson grants that stories like *Alice in Wonderland* can be unnatural, unnatural narrative theory is more interested in postmodernist texts and less in speculative fiction. Jan Alber elaborates that the impossibilities encountered in for instance fantasy and science fiction are conventionalized: they are genres in which impossibilities can be explained through magic or the supernatural, or in which similar phenomena are explained through technological development.[19] Weird fiction appears to oscillate somewhere in-between, as speculative fiction that does adhere to certain quite specific genre or subcultural conventions (for instance the tentacle as a symbol of the weird).[20] At the same time, though, the weird violates expectations and conventions in order to achieve the destabilization that unnatural narratives demand.

The main claim in this chapter is that contemporary narratives using the weird mode are in a good position to activate the reader's environmental imagination *because of* the impossible spaces presented and the ways in which they are presented. This claim also corresponds well with Alber's description of unnatural scenarios and events as "particularly well designed to make us more flexible because such textual segments urge us to deal with extremely recalcitrant concepts"—concepts such as the Anthropocene.[21] With this in mind, however, the weird is marked by a "slipperiness of form, a refusal to fit narrative or generic expectation."[22] Instead of placing weird story-spaces too firmly in any narrative theory, therefore, "the weird" as a liminal threshold notion might find a more "natural" connection to theory and ecocriticism via Morton's ecological thought and the Anthropocene discourse more broadly. As Morton claims, the Anthropocene—also a so-called "threshold concept"[23]—"brings together human history and geological time in a strange loop, weirdly weird."[24]

Roger Luckhurst also ties the weird to ecological concerns via its "strange territorial elasticity of borderscapes" consisting of "temporary heterotopic zones" that imply "movement between the real and the unreal."[25] Lovecraft's metaphor of the threshold as an access point to the weird is therefore a useful figure to describe weird spatiality more generally. The weird, as Fisher notes, always "presents us with a threshold between worlds."[26] This threshold usually involves intrusions from a supernatural, "weird" outside into a perceived natural, "normal" inside—sometimes instigated by the hapless protagonists of weird tales, who must forever contemplate the existence of this terrible outside. Often the intrusion is related second-hand, like in "The Colour Out of Space" and "The Call of Cthulhu," via one or more frame narratives. This implied physical and experiential distance between the narrator and the weird is used to emphasize the elusiveness of the creeping horror experienced by the narrator, and so the narrative situation in weird stories can reflect the threshold they thematize; other examples from the traditional weird are Algernon Blackwood's *John Silence* stories (1908) and William Hope Hodgson's *The House on the Borderland* (1908). Newer weird stories like Mark Danielewski's *House of Leaves* (2000) and VanderMeer's Southern Reach trilogy probe the agency of the threshold itself, and ask what happens when the threshold expands beyond simple oppositions between outside and inside, known and unknown.

Area X and the Brave New Weird

The Southern Reach trilogy (*Annihilation, Authority, Acceptance*, all of which were originally—and successively—published in 2014) follows a set of characters through their figurative and literal absorption into the mysterious zone of pristine wilderness called "Area X." The Southern Reach, a government research facility controlled by the elusive security service

"Central," is positioned near the strange border of Area X and sends increasingly desperate expeditions across to identify its origins and purpose. In the course of *Annihilation* (set inside Area X) and *Authority* (set at the Southern Reach institute), it becomes evident that Area X is in the process of assimilating and altering everything it touches. At the end of *Acceptance*, focalized from interchanging character perspectives, the influence of Area X has become an irreversible part of reality: "Area X was all around them; Area X was contained in no place or figure. [...] It was the heavens and earth," and "the hegemony of what was real had been broken, forever."[27]

The ecological outlook of the trilogy has, since its publication in 2014, inspired a number of scholars to engage with VanderMeer's storyworld, using new genre-markers like "weird ecology"[28] "ecological uncanny"[29] and "dark ecology."[30] The trilogy has been discussed in relation to philosophy such as Heidegger's metaphysics,[31] object-oriented ontology[32] and posthumanism,[33] and considered in terms of literary-theoretical concepts like econarratology[34] and ecosickness.[35] In *None of This Is Normal* (2018), the first monograph on VanderMeer's work, Benjamin Robertson reads Area X geopolitically, representing a "fantastic materiality" which contrasts Lovecraft by disrupting borders "between that and this, between one space or time and another space or time, between the human and whatever its other happens to be."[36]

VanderMeer has, similar to Lovecraft, modeled Area X after a place he knows well: the Saint Marks wildlife refuge in Florida.[37] Like Lovecraft, VanderMeer uses the weird as a mode to express spatial or cosmic anxieties related to the unknown and the nonhuman. Tabas notes that VanderMeer, like Lovecraft, displays a weird realism which "recalls traditional nature writing in its content and its form [yet] seems to look *too* closely at things, to discover their weirdness and terror, not their familiarity and presence" through obsessive spatial description.[38] For all these similarities, however, VanderMeer's prose displays qualities that depart from Lovecraft and the traditional weird. By using narrative techniques such as obsessive spatial description and the deliberate failure thereof, but mediated through unconventional characterization and focalization, VanderMeer approaches Anthropocene uncertainties by leaning more toward Solnit's proposed hopeful unknown and affirmative thinking than toward Lovecraft's fearful unknown and nihilism.

Of the trilogy, *Annihilation* especially recalls a sense of absorption into pristine nature, as the narrator known only as "the biologist" wanders through the weird environment of Area X:

> The richness of Area X's biosphere was reflected in the wealth of birdlife, from warblers and flickers to cormorants and black ibis. I could also see a bit into the salt marshes, and my attention there was rewarded by a minute-long glimpse of a pair of otters. At one point, they glanced up and I had a strange sensation that they could see me watching them. It was a feeling I often had when out in the wilderness: that things were not quite what they seemed, and I had to fight against the sensation because it could overwhelm my scientific objectivity.[39]

Comparable passages of careful environmental observation run through VanderMeer's trilogy, communicating an acute spatial awareness central to weird prose. *Annihilation* is the biologist's journal report of the expedition, which adds a layer of accuracy, as does the knowledge that she is an expert on transitional ecosystems. As readers, we are more likely to consider the biologist reliable, and thus also more likely to accept the weird reality presented, because she exhibits such level-headed reflection. For example, the biologist describes the sudden appearance of dolphins in the river as "a dislocation" because she, already attuned to the weirdness of Area X, had expected something *less* explicable: "when the mind expects a certain range of possibilities, any explanation that falls outside of that expectation can surprise."[40] Likewise, the nearly instantaneous reversal back to weird as the biologist notices that the eye of the dolphin is "painfully human, almost familiar" becomes deeply uncanny because the biologist has explained the scene in such a realistic manner. The effect of the close, empirically voiced study of setting is that Area X becomes unnerving not, primarily, because it is perceived as weird, but because it is perceived as *real*.

Fisher notes, recalling unnatural narrative, that the weird "involves a sensation of *wrongness*: a weird entity or object is so strange that it makes us feel that it should not exist, or at least it should not exist here."[41] Yet, Fisher goes on, "the weird cannot only repel, it must also compel our attention."[42] The opening sentences of *Annihilation* perform Fisher's definition:

> The tower, which was not supposed to be there, plunges into the earth in a place where the black pine forest begins to make way to swamp and then the reeds and wind-gnarled trees of the marsh flats. Beyond the marsh flats and the natural canals lies the ocean and, a little farther down the coast, a derelict lighthouse.[43]

How is the reader supposed to respond to a tower, which is *not* supposed to be there, and exhibits behavior contrary to most towers as it plunges *into* the earth? How can the reader negotiate these two empirically flawed observations with the perfectly rational, elaborate description of the landscape surrounding the *not*-tower? Bewildering topography and impossible objects cut into the obsessive environmental description, leaving ruptures that elude satisfactory explanation just as they demand further investigation. This is how the weird maximizes a sense of wrongness, a sense of spatial or existential discordance. This spatial discordance can encourage the reader to move beyond mere wrongness and dare to engage with the ruptures created rather than succumb to paralysis or nihilism.

As noted above, VanderMeer and Lovecraft both make use of obsessive spatial description, although in different ways. Relating to this meticulous anti-/realism of the weird, moreover, Harman and Kneale (citing Harman) both point out Lovecraft's use of *catachresis*: which they define as the deliberate debilitation of the powers of language.[44] Lovecraft employs catachresis as a tool to increase the horror of the unknown. For example, the conspicuous nondescription of the weird phenomenon in "The Colour Out of Space" complements the careful, almost rhythmic environmental description in the rest of the story and suggests that the elusiveness of the weird is reason enough to fear it. Lovecraft's "Colour" in fact bears similarities to *The Southern Reach*, as it features an alien life form that resides in a well outside the fictional Arkham, having arrived per meteorite. The large area around the crater is called the "blasted heath," without vegetation and covered by "only a fine grey dust or ash which no wind seemed ever to blow about."[45] Like Area X, the blasted heath is the source of ecological weirding, but it is more malignant in how it absorbs ecological life around it. More importantly, however, the description of the Colour is excessive, frantic, yet fails to address the entity directly: an "alien and undimensioned rainbow of cryptic poison from the well—seething, feeling, lapping, reaching, scintillating, straining, and malignly bubbling in its cosmic and unrecognizable chromaticism."[46] This passage contains an elaborate description that decomposes into sensory fragments, all of which fail to explain the whole. S. T. Joshi mentions this passage to accredit Lovecraft's artistic restraint, and Harman, Kneale, and Tabas quote other similar examples in their discussion of weird style.[47]

Mark Blacklock, likewise, suggests that Lovecraft's "crisis of representation set in chain by non-Euclidean and n-dimensional geometries" can be read in terms of "higher spatial form": Lovecraft's crisis of representation is a deliberate narrative strategy used "to confront his readers with unknown and dreadful forces."[48] But a story like "The Colour Out of Space" tends to stay on one side of the threshold, as it were, looking *out* toward these unknown and dreadful forces and suggesting either nihilistic surrender or a masochistic desire to remain eternally humbled and awed by them. "The Colour," like so many of Lovecraft's tales, ends with the suggestion that the "black extra-cosmic gulfs" are utterly indifferent to human quandaries, and will continue to demand our attention regardless of how we feel about it:

> It was just a colour out of space—a frightful messenger from unformed realms of infinity beyond all Nature as we know it; from realms whose mere existence stuns the brain and numbs us with the black extra-cosmic gulfs it throws open before our frenzied eyes.[49]

There is a sense of finality in the surrender above, leaving little space for uncertainty, and therefore little space for hope, according to Solnit's definition quoted above. The effect of Lovecraft's catachresis on a stylistic level is therefore limited by his usual endgame on plot level: the culmination in a nihilist philosophical view which implies or declares that comprehension of the weird is by definition impossible, and accepting this leads, by default, to eternal discomfort and terror. Consider, for comparison, another passage from *Annihilation*, in which the biologist encounters the monstrous "Crawler," the driving force of Area X's eco-colonization:

> It was a figure within a series of refracted panes of glass. It was a series of layers in the shape of an archway. It was a great sluglike monster ringed by satellites of even odder creatures. It was a glistening star. My eyes kept glancing off of it as if an optic nerve was not enough. ... The shape spread until it was even where it was not, or *should not have been*.[50]

The biologist expresses herself using the same detailed, allusive language as in her landscape account, ending once again in the spatial insistence of something which "should not have been." But she must resort to layers of sensory description, the narrative pace speeding up with each fragmented association, failing to convey her experience. As in Lovecraft, this catachrestic tension builds up to manifest in the ineffable. But where Lovecraft ends, VanderMeer pushes further. The biologist experiments with various ways of tricking her senses in order to perceive the Crawler, but admitting that "this moment of an encounter with the most beautiful, the most terrible thing I might ever experience—was beyond me," she is left "at the threshold, watching the Crawler, frozen."[51] Again, the metaphor of a threshold is used to describe the separation of the weird from the individual, but instead of recoiling in fear or madness, fascination takes over and tips the scale of ineffability from cosmic dread to the sublime.

The most interesting move VanderMeer makes in this scene, however, is asking explicitly what comes after the ineffable is a fact. The biologist wonders: "I might have watched it forever and never noticed the awful passage of the years. But then what? What occurs after revelation and paralysis?"[52] In Lovecraft, the answer would be madness or death. The biologist, however, gives a third option: "Either death or a slow and certain thawing."[53] She thaws from her paralysis, engages with the Crawler, and after enduring excruciating pain from the contact, she reemerges from the tunneling tower, reflecting on Area X's unrelenting colonization of Earth: "The terrible thing, the thought I cannot dislodge after all I have seen, is that I can no longer say with conviction that this is a bad thing."[54] Challenging catachresis can bring catharsis, VanderMeer's weird seems to suggest. The biologist, through her affinity for observing ecosystems, her factual description, and subsequent coping with catachresis, is in a perfect position to "trick" the reader into engaging with Area X. As a voice channeling ecological thought, therefore, the perspective of the biologist pushes past paralysis and invites the reader to do the same.

From the beginning the biologist shows an affinity for understanding Area X that differs from the other expedition members, most clearly manifest in her perception of the "topographical anomaly" they encounter as a tower instead of a tunnel: "Something about the idea of a tower that headed straight down played with a twinned sensation of vertigo and fascination with structure. I could not tell which part I craved and which part I feared."[55] Through the first-person perspective, the reader is invited to identify with and think like the biologist, which, as noted above, increases the effect of the weird interrupting her spatial focus. *Authority* employs a more typical past-tense, third-person narration, focalized internally through Control (a nickname that becomes increasingly ironic throughout the story), and interspersed with a few present-tense dreams and some free indirect discourse specifically in the more horrific encounters with the weird. In *Acceptance*, five perspectives and narrative timelines (pre-, during-, and post-Area X) mingle, creating a sense of temporal and spatial enmeshment that reflects the characters' absorption into Area X. The most interesting of the five is arguably Ghost Bird: Area X's copy of the biologist.

Ghost Bird is focalized internally through a third-person narrator. She is an Area X creation and emissary with a body and mind that consists of human and nonhuman parts. Because of the human-nonhuman merging via a third-person narrator, Ghost Bird, as a focalizer, dances on the edge between defamiliarization and empathy. This dialectic, Lars Bernaerts and his co-writers argue, is typical for nonhuman narrators, "unsettling because they blur the boundary between the human and the non-human by asking their readers to imaginatively adopt perspectives radically different from their own everyday experience."[56] In VanderMeer's chimeric character, biological ties to humans are still present, thus inviting the reader to enter a mindscape that is relatable to a *form* of humanity. However, the anthropomorphic Ghost Bird is also radically different from humanity in ways that question a conventional understanding of subjectivity and agency.

The mental distance between Ghost Bird and the other human characters is communicated via another threshold-like metaphor: she feels as if she is "looking at them from the border [of Area X]—peering in through the shimmering door."[57] The implication is that Ghost Bird is on the *other* side of the threshold, looking in on the human world. At one point in *Acceptance*, Ghost Bird criticizes the human way of thinking in terms of linear, goal-oriented purpose: "As if *purpose* could solve everything, could take the outlines of what was missing and by sheer will invoke it, make it appear, bring it back to life."[58] This suggests a desire to explore other coordinates by which more-than-human considerations may also be included in the mapping of reality. Through Ghost Bird's gaze, VanderMeer thus invites the reader to take on a nonhuman perspective, and challenges the stability of the category of "human"—even more so because the character is essentially an extension of Area X.

Area X itself takes on an unsettling agency, most directly via Ghost Bird's perspective, but also through the contamination of the other characters: they all refer to the "inquisitive brightness" of Area X that takes control of their bodies and minds.[59] Ghost Bird, especially, reveals the alarming sentience of the ecosystem: "Area X was analyzing her from all sides. It made her feel like an outline created by the regard bearing down on her, one that moved only because the regard moved with her, held her constituent atoms in a coherent shape."[60] Marco Caracciolo calls this an "actantial mediation of place" and argues that "the landscape of Area X thus ends up taking over and nonhumanizing the human."[61] In a sense, VanderMeer's trilogy acts out one of the much-discussed feats of the Anthropocene, what Latour calls a "surprising inversion of background and foreground," where nature becomes the acting subject and humanity the passive object.[62] From the changing perspectives in *Acceptance*, it might even be suggested that Area X has taken on the role of a barely visible, omniscient narrator, one that haunts the characters—and by extension the readers—to varying degrees, depending on their level of contamination.

The two chapters framing *Acceptance* are narrated from the jarring second-person perspective of Gloria, former director of the Southern Reach institute, as she is dying inside Area X. She describes an "interrogation going on" by "a kind of alien regard," which "will repeat until you have given up every answer."[63] In the last chapter, Gloria feels that "Area X is done with you, has taken every last little thing out of you."[64] The implication is that Area X actively "interrogates" or analyzes the life forms it encounters in order to assimilate the environment to its design. This idea is reinforced by Ghost Bird, who tries to explain to Control Area X's complete dominance:

> Have you not understood yet that whatever's causing this can manipulate the genome, works miracles of mimicry and biology? Knows what to do with molecules and membranes, can *peer through things*, can surveil, and then withdraw. [...] that it's operating off of such refined and intricate senses that the tools we've bound ourselves with, the ways we record the universe, are probably evidence of our own primitive nature.[65]

This passage, focalized through Control, primes the reader for the question Control asks himself next, breaking the narration with a short, italicized inner monologue: "*Is there something in the corner of your eye that you cannot get out?*"[66] The comment can be read in two ways: as Control reacting to something in the landscape, or as Area X addressing the reader directly. The latter suggests, rather disturbingly, that Area X truly "peers through" *everything*, its control extending beyond the plot to the level of narration.

Tabas notes that the disorientation found in Area X "springs not from an absence of reality, but from an excess of the real."[67] The insidious presence of Area X as a diegetic agent is mirrored in its strong spatial presence at plot level, as discussed above. Area X's intensely real spatiality is at odds with its reluctance to follow the laws of nature, creating Escheresque spatial ruptures that VanderMeer's characters are forced to engage with even as their "five senses are not enough."[68] Read in terms of unnatural narrative, therefore, "by taking us to the most remote territories of what can be imagined," the Southern Reach trilogy and weird stories like it can "significantly widen the cognitive horizon of human awareness."[69] Area X is more than an outside force interrupting and looming threateningly in the background of delicate human reality. Area X's ecosystem physically interferes at all levels of human and nonhuman life, thus interconnecting and transforming the entire earth system. This organic enmeshment, reinforced structurally by the actantial mediation of Area X, suggests a desire to rewire the fabric of reality rather than a resignation to remain in a stable opposition between outside and inside, cosmic and local, weird and natural. In *Acceptance*, Ghost Bird explicitly voices this desire: "Perhaps a copy could also be superior to the original, create a new reality by avoiding old mistakes."[70]

CONCLUSION

Anthropocene discourse often gives a sense of apocalyptic urgency *and* of utter incapability to tie this urgency to anything easily comprehensible, which becomes frustrating and paralysis-inducing. Compared to the spatial dissonance at the heart of weird narrative, the tension between crisis and paralysis suggests one possible reason why the weird appears to be well-suited for expressing environmental anxieties. This chapter has suggested that recent or New Weird narrative demonstrates both iteration and evolution of anxieties explored in traditional or Old Weird, and identified some of the techniques used to express those anxieties. Weird prose contains a paradoxical obsession with and simultaneous revulsion by the unknown, manifest in the spatial focus and anti-mimetic structure reminiscent of, but not necessarily identical with, unnatural narrative. Moreover, stories like *The Southern Reach* express and digest contemporary ecological anxieties by experimenting with narrative perspectives that disturb conventional conceptions of ecology and the position of humans therein. This experimental perspective-taking is also where VanderMeer shows most clearly that the New Weird has departed from the Old Weird, as the perspectives conjure a strangely hopeful tone. "Hope," Solnit writes, "is "the belief that what we do matters even though how and when it may matter, who and what it may impact, are not things we can know beforehand."[71]

At the end of the third book, Ghost Bird expresses a wish to observe the rest of the world "through the eyes of Area X."[72] Considering Area X's uncanny omniscience, it is tempting to understand the very act of reading *The Southern Reach* as through the eyes of Area X. "It was just an ordinary summer day," Ghost Bird states in the last scene. "So they walked forward, throwing pebbles as they went, throwing pebbles to find the invisible outline of a border that might not exist anymore. They walked for a long time, throwing pebbles at the air."[73] Perhaps the title *Acceptance* gives a hint as to how readers could engage with VanderMeer's trilogy, accepting the spatial discordance and the weird agency of Area X in the same way that Solnit urges us to embrace the unknown. Accepting VanderMeer's unknown is taking on Area X's perspective, and (dis)locating hope somewhere in the future. It is a slippery, displaced hope, which involves considering that Area X—the world *after us*—is not necessarily better or worse, because it depends on who "us" is. Area X is ordinary, different, *in*different, and very, weirdly real.

NOTES

1. Mark Fisher, *The Weird and the Eerie* (London: Repeater, 2016), 16.
2. See Graham Harman, *Weird Realism: Lovecraft and Philosophy* (Winchester: Zero, 2012); Brian Johnson, "Prehistories of Posthumanism," and Patricia MacCormack, "Lovecraft's Cosmic Ethics," both in *The Age of Lovecraft*, eds. Carl H. Sederholm and Jeffrey A. Weinstock (Minneapolis, MA: University of Minnesota Press, 2016).
3. The Anthropocene is the name suggested by a group geologists, since adopted throughout the humanities and sciences, for the current planetary age marked by (detrimental) human activity. See Paul Crutzen and Eugene F. Stoermer, "The 'Anthropocene,'" *Global Change Newsletter* 41 (2000): 41–42, IGBP.
4. Brad Tabas, "Dark Places: Ecology, Place, and the Metaphysics of Horror Fiction," *Miranda* 11 (July 2015): 1–17, http://doi.org/10.4000/miranda.7012.
5. Fisher, *The Weird and the Eerie*, 20; see also Harman, *Weird Realism*.
6. James Kneale, "Ghoulish Dialogues," in *The Age of Lovecraft*, eds. Sederholm and Weinstock, 44.
7. MacCormack, "Lovecraft's Cosmic," 213.
8. Alexa Weik von Mossner, *Affective Ecologies: Empathy, Emotion, and Environmental Narrative* (Ohio State University Press, 2017), 163.
9. Benjamin Noys and Timothy S. Murphy, "Introduction: Old and New Weird," *Genre* 49, no. 2 (July 2016): 125, http://doi.org/10.1215/00166928-3512285.
10. Erin James, *The Storyworld Accord: Econarratology and Postcolonial Narratives* (Lincoln: University of Nebraska Press, 2015), 33.
11. Timothy Morton, *The Ecological Thought* (Cambridge, MA: Harvard University Press, 2010), 17.
12. H.P. Lovecraft, "Supernatural Horror in Literature," 1927, in *H. P. Lovecraft: The Complete Fiction* (New York, NY: Barnes & Noble, 2011), 1043, 1041–98.

13. Rebecca Solnit, *Hope in the Dark* (Chicago, IL: Haymarket Books, 2016), xiv.
14. H.P. Lovecraft, quoted in S. T. Joshi, *The Rise, Fall and Rise of the Cthulhu Mythos* (New York, NY: Hippocampus Press, 2015), 17.
15. Joshi, *The Rise*, 17.
16. Fisher, *The Weird and the Eerie*, 20, 24.
17. Marie-Laure Ryan, "Space," in *The Handbook of Narratology* (Berlin: Walter de Gruyter, 2009), 430.
18. Brian Richardson, "Unnatural Narratology," *Diegisis* 1, no. 1 (2012): 95, http://d-nb.info/1028308531/34.
19. Jan Alber, *Unnatural Narrative: Impossible Worlds in Fiction and Drama* (Lincoln: University of Nebraska Press, 2016), 145, 182, 212.
20. China Miéville, "Weird Fiction," in *The Routledge Companion to Science Fiction*, eds. Mark Bould et al. (London: Routledge, 2009), 510–16.
21. Alber, *Unnatural*, 216.
22. Roger Luckhurst, "The Weird: A Dis/Orientation," *Textual Practice* 31, no. 6 (2017): 1050, https://doi.org/10.1080/0950236X.2017.1358690.
23. Timothy Clark, *Ecocriticism on the Edge* (London: Bloomsbury, 2015), 9.
24. Morton, *Dark Ecology* (London: Verso, 2016), 8.
25. Luckhurst, "The Weird," 1060.
26. Fisher, *The Weird and the Eerie*, 18.
27. Jeff VanderMeer, *Acceptance* (London: Fourth Estate 2015), 283, 329.
28. David Tompkins, "Weird Ecology: On the Southern Reach Trilogy," *LA Review of Books*, September 30, 2014, https://lareviewofbooks.org/article/weird-ecology-southern-reach-trilogy/.
29. Siobhan Carrol "The Ecological Uncanny: On the 'Southern Reach' Trilogy," *LA Review of Books*, October 5, 2015, https://lareviewofbooks.org/article/the-ecological-uncanny-on-the-southern-reach-trilogy/.
30. Andrew Hageman, "A Conversation Between Timothy Morton and Jeff VanderMeer," in *Global Weirding*, eds. Gerry Canavan and Andrew Hageman (Vashon Island, WA: Paradoxa, 2016).
31. Tabas, "Dark Places," 14–15.
32. Matthew Masucci, "Angry Eden," in *Dark Nature: Anti-Pastoral Essays in American Literature and Culture* (Lanham: Lexington Books, 2016), 171.
33. Tom Idema, *Stages of Transmutation: Science Fiction, Biology, and Environmental Posthumanism* (New York, NY: Routledge, 2018).
34. Marco Caracciolo, "Notes for an Econarratological Reading of Character," *Frontiers of Narrative Studies* 4, no. 1 (2018), 184–85.
35. Alison Sperling, "Second Skins: A Body-Ecology of Jeff VanderMeer's *The Southern Reach Trilogy*," in *Global Weirding*, eds. Gerry Canavan and Andrew Hageman (Vashon Island, WA: Paradoxa, 2016), 220. See also Heather Houser, *Ecosickness* (New York, NY: Columbia University Press, 2014).

36. Benjamin Robertson, *None of This Is Normal* (Minneapolis, MN: University of Minnesota Press, 2018), 116.
37. VanderMeer, "From *Annihilation* to *Acceptance*: A Writer's Surreal Journey," *The Atlantic*, January 28, 2015, www.theatlantic.com/entertainment/archive/2015/01/from-annihilation-to-acceptance-a-writers-surreal-journey/384884/.
38. Brad Tabas, "Dark Places," 11.
39. VanderMeer, *Annihilation* (London: Fourth Estate, 2014), 30.
40. Ibid., 97.
41. Fisher, *The Weird and the Eerie*, 15.
42. Ibid., 19.
43. VanderMeer, *Annihilation*, 3.
44. Harman, *Weird Realism*, 42; Kneale, "Ghoulish," 46.
45. Lovecraft, "The Colour Out of Space," 1927, in *Tales of H. P. Lovecraft*, ed. Joyce Carol Oates (New York, NY: Harper Perennial, 2011), 78.
46. Lovecraft, "The Colour," 97.
47. Joshi, *The Rise*, 97.
48. Mark Blacklock, "Higher Spatial Form in Weird Fiction," *Textual Practice* 31, no. 6 (2017): 1111, https://doi.org/10.1080/0950236X.2017.1358690.
49. Lovecraft, "The Colour," 99.
50. VanderMeer, *Annihilation*, 177.
51. Ibid., 178–79.
52. Ibid., 179.
53. Ibid.
54. Ibid., 192.
55. Ibid., 14.
56. Lars Bernaerts et al., "The Storied Lives of Nonhuman Narrators," *Narrative* 22, no. 1 (2014), 25, http://hdl.handle.net/1854/LU-4254459.
57. VanderMeer, *Acceptance*, 192.
58. Ibid., 191.
59. VanderMeer, *Authority*, 262.
60. VanderMeer, *Acceptance*, 37.
61. Caracciolo, "Notes," 185.
62. Bruno Latour, "Agency at the time of the Anthropocene," *New Literary History* 45 (2014), *Project MUSE*, 14, http://doi.org/10.1353/nlh.2014.0003.
63. VanderMeer, *Acceptance*, 7–8.
64. Ibid., 337.
65. Ibid., 81.
66. Ibid.
67. Tabas, "Dark Places," 12.
68. VanderMeer, *Annihilation*, 178.

69. Alber, *Unnatural*, 216.
70. VanderMeer, *Acceptance*, 35.
71. Rebecca Solnit, *Hope*, xiv.
72. VanderMeer, *Acceptance*, 329.
73. Ibid., 331.

BIBLIOGRAPHY

Alber, Jan. *Unnatural Narrative: Impossible Worlds in Fiction and Drama*. Lincoln: University of Nebraska Press, 2016.

Blacklock, Mark. "Higher Spatial Form in Weird Fiction." *Textual Practice* 31, no. 6 (2017): 1041–61.

Caracciolo, Marco. "Notes for an Econarratological Reading of Character." *Frontiers of Narrative Studies* 4, no. 1 (2018): 172–89.

Carrol, Siobhan. "The Ecological Uncanny: On the 'Southern Reach' Trilogy." *LA Review of Books*, October 5, 2015. https://lareviewofbooks.org/article/the-ecological-uncanny-on-the-southern-reach-trilogy/.

Clark, Timothy. *Ecocriticism on the Edge*. London: Bloomsbury, 2015.

Crutzen, Paul, and Eugene F. Stoermer. "The 'Anthropocene'." *Global Change Newsletter* 41 (2000): 41–42. *IGBP*. www.igbp.net/.

Fisher, Mark. *The Weird and the Eerie*. London: Repeater, 2016.

Hageman, Andrew. "A Conversation Between Timothy Morton and Jeff Vander-Meer." In *Global Weirding*, edited by Gerry Canavan and Andrew Hageman, 41–65. Vashon Island, WA: Paradoxa, 2016.

Harman, Graham. *Weird Realism: Lovecraft and Philosophy*. Winchester, UK: Zero Books, 2012.

Houser, Heather. *Ecosickness*. New York, NY: Columbia University Press, 2014.

Idema, Tom. *Stages of Transmutation: Science Fiction, Biology, and Environmental Posthumanism*. New York, NY: Routledge, 2018.

James, Erin. *The Storyworld Accord: Econarratology and Postcolonial Narratives*. Lincoln: University of Nebraska Press, 2015.

Johnson, Brian. "Prehistories of Posthumanism." In *The Age of Lovecraft*, edited by Carl H. Sederholm and Jeffrey A. Weinstock, 97–116. Minneapolis, MA: University of Minnesota Press, 2016.

Joshi, S.T. *The Rise, Fall and Rise of the Cthulhu Mythos*. New York, NY: Hippocampus Press, 2015.

Kneale, James. "Ghoulish Dialogues: H. P. Lovecraft's Weird Geographies." In *Age of Lovecraft*, edited by Carl H. Sederholm and Jeffrey A. Weinstock, 43–61. Minneapolis, MA: University of Minnesota Press, 2016.

Latour, Bruno. "Agency at the Time of the Anthropocene." *New Literary History* 45 (2014): 1–18.

Lovecraft, H.P. "Supernatural Horror in Literature." 1927. In *H. P. Lovecraft: The Complete Fiction*, 1041–98. New York, NY: Barnes & Noble, 2011.

———. "The Colour Out of Space." 1927. In *Tales of H. P. Lovecraft*, edited by Joyce Carol Oates, 77–100. New York, NY: Harper Perennial, 2011.

Luckhurst, Roger. "The Weird: A Dis/Orientation." *Textual Practice* 31, no. 6 (2017): 1041–61.

MacCormack, Patricia. "Lovecraft's Cosmic Ethics." In *The Age of Lovecraft*, edited by Carl H. Sederholm and Jeffrey A. Weinstock, 199–214. Minneapolis, MA: University of Minnesota Press, 2016.

Masucci, Matthew. "Angry Eden: Hyperobjects, Plant Entelechy, and the Horror of Eco-Colonization in Jeff VanderMeer's Southern Reach Trilogy." In *Dark Nature: Anti-Pastoral Essays in American Literature and Culture*, edited by Richard J. Schneider, 171–84. Lanham: Lexington Books, 2016.

Miéville, China. "Weird Fiction." In *The Routledge Companion to Science Fiction*, edited by Mark Bould, Andrew M. Butler, Adam Roberts, and Sherryl Vint, 510–16. London: Routledge, 2009.

Morton, Timothy. *Dark Ecology*. London: Verso, 2016.

———. *The Ecological Thought*. Cambridge, MA: Harvard University Press, 2010.

Noys, Benjamin, and Timothy S. Murphy. "Introduction: Old and New Weird." *Genre* 49, no. 2 (July 2016): 117–34. http://doi.org/10.1215/00166928-3512285.

Richardson, Brian. "Unnatural Narratology." *Diegisis* 1, no. 1 (2012): 95–103. http://d-nb.info/1028308531/34.

Robertson, Benjamin. *None of This Is Normal*. Minneapolis, MN: University of Minnesota Press, 2018.

Ryan, Marie-Laure. "Space." In *The Handbook of Narratology*, edited by Peter Hühn, John Pier, Wolf Schmid, and Jörg Schönert, 420–33. Berlin: Walter de Gruyter, 2009.

Solnit, Rebecca. *Hope in the Dark*. Chicago, IL: Haymarket Books 2016.

Sperling, Alison. "Second Skins: A Body-Ecology of Jeff VanderMeer's *The Southern Reach Trilogy*." In Global Weirding, edited by Gerry Canavan and Andrew Hageman, 241–68. Vashon Island, WA: Paradoxa, 2016.

Tabas, Brad. "Dark Places: Ecology, Place, and the Metaphysics of Horror Fiction." *Miranda* 11 (July 2015): 1–17. http://doi.org/10.4000/miranda.7012.

Tompkins, David. "Weird Ecology: On the Southern Reach Trilogy." *LA Review of Books*, September 30, 2014. https://lareviewofbooks.org/article/weird-ecology-southern-reach-trilogy/.

VanderMeer, Jeff. *Acceptance*. London: Fourth Estate, 2014.

———. *Annihilation*. London: Fourth Estate, 2014.

———. *Authority*. London: Fourth Estate, 2014.

———. "From *Annihilation* to *Acceptance*: A Writer's Surreal Journey." *The Atlantic*, January 28, 2015. Accessed May 16, 2019.www.theatlantic.com/

entertainment/archive/2015/01/from-annihilation-to-acceptance-a-writers-surreal-journey/384884/.

Weik von Mossner, Alexa. *Affective Ecologies: Empathy, Emotion, and Environmental Narrative*. Columbus: Ohio State University Press, 2017.

CHAPTER 10

Inexistent Ink: Michael Cisco and Quentin Meillassoux on Writing Worlds

Ben Woodard

From *The Divinity Student* (1999) to *Unlanguage* (2018) Michael Cisco's articulation of the weird touches on the oblique construction that accompanies the narrative matter of text itself (how what is written accounts for the effect of being read). But rather than discussing written marks as a material affect, the matter of inscription will be analyzed as an imperfect index of another world (whether actual or possible). By inscription here, I mean the material generation of a sign that is mean to cause structural change in a thinker. This analysis will not focus necessarily on terms of the imaginary or the fantastic but think in terms of indexing the proper world structure of each inscription suggesting less fantastical realms than formally inexistent places. Thus, Cisco's weird worlds dwell somewhere between H. P. Lovecraft's material possibilities (without escaping the ideal) and Thomas Ligotti's psychological impossibilities (without descending into cosmically constitutive madness).

B. Woodard (✉)
Leuphana University of Lüneburg, Lüneburg, Germany

© The Author(s) 2019
J. Greve and F. Zappe (eds.), *Spaces and Fictions of the Weird and the Fantastic*, Geocriticism and Spatial Literary Studies,
https://doi.org/10.1007/978-3-030-28116-8_10

To this end the following chapter will examine Cisco's work in relation to that of Quentin Meillassoux's discourse on the "meaningless sign" as well as the relationship he draws between inscription and the radically contingent nature of reality in contradiction to the relative stability of any (fictional) world within it. If anything can be written (and anything can happen), how do we understand the limits of writing in terms of the limits of consciousness (and the thinkability of the concept of a world which is a constellation of limits and conditions)?

Whereas the better known discourse on the materiality of writing (Blanchot, Derrida) suggests the power of writing to capture something of the impossible or the incomplete, Meillassoux claims that a certain type of writing expresses the absolute chaos of the world in its simple banality. In a similar sense, Cisco's purposefully uneven absorption (or cautious usurpation) of the formal constraints of narration, structure, and language create a hallucinatory metafiction that degrades the split between the grasp of the story (on the page) and the telling of the story (off the page). This hallucinatory reading positions the creative mind as either an expression of the experience of contingency ontologically loaded (Deleuze) or rationality's grasp of the trace of absolute contingency (Meillassoux). The very space of this difference between Gilles Deleuze and Meillassoux indexes the concept of writing from the *atmosphere itself* as Cisco attempts to do.

Writing the Body: *The Divinity Student*

Michael Cisco's work is notoriously difficult to describe. Even in the broad categorizations of genre it falls in the cracks between gothic, fantasy, weird, and "proper" literary fiction. Jeff VanderMeer once referred to Cisco (and not in a hyperbolic way) as the American Kafka.[1] There is something telling about this. It speaks to Cisco as not simply an American version of Kafka but a Kafka "doing things big," as if he had grown up in New York City or some other sprawling absurd metropolis. That is to say, Cisco seems like a Kafka that entertains the outside of literature (or of genre) not only in the testing of its limitations but in making certain aspects of the form of representation part of the world presented. Or in other words, it is not an experimentalism that tends toward the transgression of boundaries but toward inappropriate integration not only of genre but also of writing, reading, and the world separate from them.

For instance, Cisco's first novel *The Divinity Student* (1999) opens with the spectacular death of its main character who is then promptly (but oddly) resurrected in the following manner:

> Quickly they bring him inside, lay him across two sawhorses and start cutting at him—they gut him like a fish, cut open from throat to waist, red hands pull his ribs apart, head and shoulders hanging down, his arms lying flat on the ground, tugged back and forth as they empty him out. They dump his contents cooked and steaming on the floor, and bring up stacks of books and manila folders, tearing out pages and shuffling out sheets of paper, all covered with writing, stuffing them inside, tamping them down behind his ribs and crushing them together in his abdomen. What pages they select and what books they tear are of little importance, only that he be completely filled up with writing, to bring him back, to set him to the task. Then they suture him shut again—drag him to the tub (his arms and legs dangling and catching on things overturning tables and chairs) and dump him in the water, slopping blue water on gray stone pavings, and together they draw breath and drop open their mouths, screaming noiselessly as they shove his face under the running tap and pushing him full under the water with their red hands, under their wings. The Divinity Student twitches, lashing water over the lip of the tub. Gaping they push him down harder.[2]

It would be tempting, and not completely unjustified, to follow a thinker such as Maurice Blanchot or those who would tie writing to death and to the corporeal as a path into Cisco's text. So many of his texts are visceral in this way—not visceral as a description of a kind of writing but in the sense of chipping away at the line between the meat and the letter. Blanchot's writings on Kafka which talk of a chaos of misunderstanding index a great outdoors (also championed by Meillassoux) but one that seems not merely to index an impossibility as a rupture but, paraphrasing Blanchot, the impossibility of every possibility.[3]

What is constantly at play throughout Cisco's work (and what remains a question in Meillassoux) is the materiality of the inscription (which generally weighs heavily on the word materiality). The tension can be highlighted as an uneven balance between intention of the mark, the action of the mark, and the matter of the mark. Or the agency or "why" is in relation to the action or shape "how" which in turn is in relation to the "what" of the mark. There are somewhat unremarkable observations which can be made regarding the interdependency of these aspects of writing—a skilled typist (for instance) may be able to think faster while typing than when writing by hand. But this of course may re-qualify writing by hand as "more careful" or more thoughtful in an age of typing (which may easily lead to an aesthetic overqualification or where the carbon on the page is thought of as more thoughtful regardless of its content).

In some regards *The Divinity Student* attempts to suture this loose constellation of capacities into a tighter form. The titular Divinity Student sets about finding lost forgotten words, extracting them at times from the dead tissues of those who once used them. The hallucinatory or pharmacological approach of the text (and of most of Cisco's texts) is accomplished by including the means of presentation in the order of hallucination which refers to certain aspects of that mode of hallucination. Thus, it is not merely a matter of "anything goes" or cutesy metafictional reference, but of understanding the type of illusion-giving that makes the selected modes so effective.

At times, the Divinity Student encounters those who know lost words but a closer look, a deeper inspection is required one that mirrors his own resurrection quoted earlier:

> The Divinity Student—having correctly guessed that the agitation of bubbles would speed the fermentation process—watches already the yellowy ropes of Albert's memory oozing from the tissues, mixing thickly with the chemicals.... The Divinity Student is holding the jar in one hand. With a horrible face he raises it to his lips to drink, and then vomits out the window, and drinks again and vomits.[4]

It is only by also consuming a portion of Albert's life that the Divinity Student is able to "digest" Albert's memories. The direction or final end of the student's task is unclear other than as a kind of psychotic language-obsessed enlightenment but one with cultish-religious mythos (of some kind) possibly backing it. This parallels a motif present in many of Cisco's other texts to be discussed below—namely the hybridization of the fictional and the metafictional since it is the enjoyment of the text (of reading the Divinity Student) that appears to be the very stuff of which he is made. It is possible that the circular nonexplanation, that the story attempts to close its own loop that sends the reader off like a vampire looking for other texts to consume. The whole novella pushes and morbidizes the relation between the body of the text and the concept of bodies of text.

The World Is Leaking: *Member*

Cisco's novel *Member* (2013), for instance, takes the kind of well-known musings on Kafkaesque bureaucracy and metastasizes them to the level of world structure (or even cosmic possibility). Thus, rather than the world and bureaucracy achieving a gray indifference because of the latter's suffocating power, the very structure of the world (cosmologically) becomes a meaningless game.

In terms of the recognizable plot *Member* focuses on a massive planetary-scale game called Chorncendantra that is referred to as "the human game." This game does not concern only humans but also involves multiple worlds both real and artificial. The main character, Mr. Thanks, is unexpectedly recruited into the game as a courier whose task is to deliver small cans of spells and prizes to a construction site. From there the small absurdities pile up but something at bottom refuses to topple over. It part it may be because the novel starts out as a train of thought but when it stops being that or starts again becomes a challenge to discern: "I will expand the dream to engulf what surrounds me."[5] The indiscernibility between things happening and things being thought pushes Mr. Thanks to keep trying to play his part in the game "Relaxing m y mind had only brought about a causeless, meaningless sadness" even attempting to ignore the game is playing it.[6] This is the frustration of being a human in the giant system/mechanism of Chorncendantra—one knows one is a human but that this means being a small part of the human-machine but not being able to only be a part.

One of the most impressive aspects of Cisco's novel here is that feeling of being a cog in the machine does not wander into immediate and obvious existential territory. Mr. Thanks, carrying his heavy bag, is not a dreary eyed Frenchman in 1950s Paris: "'Don't imagine that you are the flaneur,' I tell myself, 'looking down on people, like you are the last human in a world of machines the passerby are all soulless robots and you're the only one who cares—that's high school shit.'"[7] Mr. Thanks is mostly just frustrated by the small things and couldn't know enough about the large things to feel so small simply because there are others that at least seem to know more (so-called operationals and high rationals)—high rationals being those creatures that "think things up."

In some regards, in *Member* Cisco could be seen as entering the territory of either William S. Burroughs or the graphic novelist Charles Burns—where one quickly leaves "this world" and enters somewhat unreal worlds (interzones and dreamspaces) that are however still attached to this world.

Burns' *X'ed* out series of graphic novels combines Burroughs' worlds with that of *Tintin* creating a kind of self-immolating orientalism that is recognizable in some of Cisco's settings (especially San Venecificio).

And yet in *Member* Cisco appears to hold the reigns tighter than this. It is the rules of the game or the world themselves that seem parasitic yet completely natural (at least to everyone but the main character). The attempt at thought to think a world only appears to add to the problem of probing the layers of rules added upon rules till the point that one is not even sure where one is and, for that matter, it is not even clear if the narrator is any more or less sure of what he is doing that what the reader is reading about him. And of course, the wonderfulness of Cisco's descriptions which are present throughout. For example:

> Perched there, he aims carefully at something I have trouble making out. It's a large, solid object that seems to be browsing along the sidewalk in its own special darkness; not a blob of shadow exactly, more like a dead, uninteresting haze of grey smoke that collects around it and projects out of it in a reverse spotlight. In overall shape, it resembles a human liver, all covered in imbricated scales. A felty, transparent caul seems to envelop the entire thing, and ripples out wrinkles and folds to palpate its surroundings, making the emitter seem both solid and liquid at once.[8]

The descriptions shift from the weird tale type above to bordering on the romantic: "In that faint, brief light, I see the tendrils of smoke from each little candle immobilized like ectoplasm calligraphy, trailing from the cake"[9] to absurd but small shifts of perspective: "Somebody left a salad out on the curb, with no bowl around it."[10] So many other images such as these exist in the text and suggest an ongoing theme of failed containment, of the rules of structures of a system never operating fully.

The question becomes less "what is happening?" and rather, "why is it that this game with its excessive rules seems normal?" The novel flirts again and again with the weirdness of games, of playing them without reasons,[11] that the problem is that we enter the game from somewhere or as something not of it,[12] that thinking and playing is just a headlong plunge into various kinds of darkness more or less familiar.[13] The novel's flatly presented insanity mirrors the very weirdness of belonging to one odd bureaucracy after another. Any attempt of pulling out seems like a childish time out: "That's what I want: a place in which I have no part. I want to ride through space like wind in wind and sleep on the void, and be a go-between with nothing but between."[14] In so much sci-fi, horror, and weird fiction there is the moment when the narrator passes into the strangeness, where the mundane becomes not the mundane, when you (as the reader) know that they are on a trajectory that ends in death, madness or, optimistically, some small share of triumph. But in *Member* Cisco's point is no such prologue exists because of the incessant nature of thought that barrier cannot be recognized—it can only be supposed after the fact when it is already too late. There is no cutting loose.

And while explicit references to philosophy appear throughout (Spinoza),[15] ancient Chinese thought, moments that smell like Heidegger and Deleuze, it would be incorrect to call this a philosophical novel (whatever that may mean) but only that, unlike many novels, *Member* makes a show of when it is pushing at its own limits or trying to catch its own tail. Every practice in the novel has its reasons but the reasons do not seem to matter other than to give you a location for your forthcoming injuries. Or, in other words, any attempt at closed precision is absurd and yet every attempt to do so is itself generative. It may be that Cisco, or Cisco's character, is having a laugh at set theory, or a system that would have its participants following the rules in a strict manner.

THE NARRATOR/UNLANGUAGE

Somewhere between *Member* and *The Divinity Student* is Cisco's *The Narrator* (originally published in 2010) in which the main character, who is a historian yet referred to as a narrator, is telling the story of the world with increasingly unexpected impacts. The narrator (an army conscript named Low) is not composed of letters and as singular in his task as the Divinity Student nor does he knows he rules of the world which is purportedly describing (contra *Member's* game player). The book is Low's narration and that which does not quite fit into since he edits and corrects descriptions (which we take to be really happening to him) as if what happened was itself only ever an account.

Much of *The Narrator* could be said to be about the horror of war and of the unreliability of history (and especially the history of war) but this unreliability is internalized to the point of becoming unrecognizable from the problem of writing after the fact of anything (horrific or otherwise). There is a potential inverse with *Unlanguage* in the sense that it is a guide to write oneself back into existence (to not be merely a person but to be a character as the opposite face of the narrator's horror at being born into character-hood). The narrator, who is the third person, is given the book of "unlanguage" (the workbook that you are told you are reading as you the read the story) and he cannot (at least for first third of the book) become a character. It is unclear, as is common with Cisco's work, what consists of the story, the book, and the difference between them in regard to what the characters have access to. While *Unlanguage* is the workbook it has narrative digressions which are referred to as miniature studies which are also accompanied by notes but whether these parables (or sub-parables) are added to the book by those reading it, as is more strongly suggested with the notes, but this is also unclear. In essence, you do not know if the narrative action is part of the book read by others and or whether the notes were added to the book by multiple different persons who read the book over time. There is a vague sense of linear narrative, but absolutely no reason to believe in one; nor to have any sense of what the difference between the characters could be (other than as first person or second person).

This indistinction itself becomes a problem for the third person. As he writes:

> I want to be a character, believe me, and speak with authority and the immortal lineaments and redundancies, but my trouble is that my 'character' is incorrigibly anti-character. The only thing that was always the same about me was that I had to be different.[16]

Furthermore, the unlanguage of the book can only be spoken and not written since it lives inside all other written languages. This of course begs the question of why having a workbook at all for a language that cannot be understood in any written form. In some regards, this is the cost of a book "about" or in unlanguage, and this is why negativity plays such an important role. The un- of unlanguage is indefinite and therefore absolute negation: "Unlanguage of unknowing expresses the absolute completely, by taking back even as it gives. This is why negation is the essence of this language, which is sometimes known as the cursed language, the language of doom."[17]

The extensive negativity (or undeadness) of the examples or parables appear increasingly in conflict with the precision of the descriptions of the function of the language especially in terms of the creativity made possible by unlanguage. One instance of this, and which sounds strikingly close to Meillassoux's notion of contingency centers on the Hypertrophaic Conditional:

What is possible cannot be represented as not existing. The possibility of a thing is that thing in an inchoacity that has the power to leap into being. So the Hypertrophaic Conditional identifies the almost-existence of a thing as an independent existence mutually exclusive of that thing. The conditionality of that thing is also only possible, because in some cases the pre-extant nebula does not transform into a thing.[18]

There is here a kind of radical plurality of existence in which every state of existence could be considered a type of existence itself (which almost seems to index Alexius Meinong more than Deleuze). But the role of intentional subjective experience is always at the forefront (and the reliance of metaphysics on this is often unacknowledged in Deleuzian scholarship, other than through a discourse of experience as supra-personal).

The degree to which a self can be said to exist or not appears central to the text and even toward the end there is an argument between a Deleuzian (the narrator of the book, the third person) and a Hegelian.[19] The body that speaks and deploys unlanguage appears trapped between neither life nor death and whose activity is expressed in terms of negation and yet as the book goes on the creative forces of the unlanguage seem to rely on ever more narrow definitions.

Again, it becomes a question of the Deleuzian status of potentiality here, of the repetition of a statement with a difference is a central part of *Unlanguage* (both the book and the language) since it is always a parasite on what is written and that it always exists within another language to some degree. But the status of this parasitism seems to rely on the intentionality or thought-directed aspect of the language and not its gesture/action nor its materiality (as a mark). For Meillassoux, as I hope to show below, certain types of gestures and actions can indicate the most basic sense of materiality, such as the trace which indexes a whole other time or world. For Meillassoux, part of the mystery is how thought is capable of recognizing and producing such traces (especially in the realm of the physical sciences or in mathematics). This is not to say that these problems are disconnected from literary writing since for Meillassoux (like for Deleuze) there are forms of writing (such as Mallarme's poetry) that seem to index an altogether different understanding or creation and of world.

WRITE THE LIGHTNING

> For all we have demonstrated, is that to produce an empty sign, one must have access to the eternity of contingency. But we have not at all shown that the empty sign allows, in turn, the description of a world independent of thought. We have only established that one must accede to eternal contingency to produce a mathematics capable of not speaking of anything—since it is founded upon meaningless signs. The new puzzle that appears before us is the following: how can a meaningless sign allow us to describe the world, without becoming once again a meaningful sign, and thereby capable of referring to a world outside of it?"[20]

> I can't return home until I've met with something eerie. The moon actually is full, and shreds of cloud really are crossing it. It would be redundant to invent poems about it because the moon is a poem. I've known that all my life. The clouds pass. I can see the moon clearly. The continents cast their shadows on it. A black jellyfish. So that's a séance going on up there, in parallel with my walk wandering down here.[21]

Both Cisco and Meillassoux bet on some capacity of mind to mine the splendorous veins of contingency in a way that is distinguishable from simply spilling out utter chaos. Meillassoux's essay "On the Meaningless Sign" attempts to update and revise his own philosophical project while building up to an argument for why the occurrence of a meaningless sign (highlighted in the quote above) not only indexes the only necessity in existence (absolute contingency) but also demonstrates the triumph of a speculative and materialist thought in the world. Broadly put, Meillassoux describes a cosmos where the task of thought in the context of a hyperchaos (where everything or nothing could happen outside the realm of possibility, i.e., with no calculable range) is to find banal invariances (to find the simplest form of thinkable consistency in a world where anything could change without reason).

For Meillassoux, mathematics is especially capable of expressing true statements about the cosmos and yet, at various points, he has stated that he believes there are truths to each domain and that the task of philosophy is to outline the primary qualities of each of these domains. Yet it is telling, and important for our discussion of Cisco, that Meillassoux goes to great length to distance himself from the philosophy of Deleuze, he who labored so extensively to saturate the sensible with thought, to claim that thought is as spontaneous an expression or imperfect capture of difference. Meillassoux sees Deleuze, and in particular the Nietzsche-sympathetic Deleuze, as the champion of subjectalism—of the dominant trend of twentieth-century thought to inject the skeleton of the cosmos with qualities of the self and the mind (whether life, will, thought, and so on). But it is important to note that Cisco is enthralled by Deleuze and, in particular, Deleuze's *Nietzsche and Philosophy* ostensibly setting him apart from Meillassoux.

How then (and why) do we bring Meillassoux's meaningless sign in relation to Cisco's work? As stated in this chapter's introduction, the peculiar way Meillassoux handles or fictionalizes the meta-functions of fiction offers a realm that produces a gap similar to the effect (at least) of Meillassoux's sign. By constantly colliding the banal and the extraordinary in a manner that makes fictional materiality of fictions conventions, Cisco's work as not belonging strictly to the literary nor to various genres weird o rfantastical, appears as a potential challenge for Meillassoux's articulation of chaos and the contingency of writing as decidedly different from that of Deleuze.[22]

Keeping the opening passage of *The Divinity Student* in mind we can examine the following striking paragraph from Meillassoux. In discussing the limit of the figural approach to speculative thought, Meillassoux admits there are certain types of facts that are domain relevant yet suggestive of a contingency beyond what is thought to be this world. The second example he gives is the following:

> The hypothetical traces of ontic fissures prohibiting a same regime of explanation from applying to different domains of the real. We will call these fissures, as 'fossil traces' of irruption *ex nihilo*, irruptive fulgurites. Fulgurites are 'lightning stones,' quasicylindrical tubes produced by the impact of lighting on a rock, or even sometimes on the sandy ground of a desert. By this metaphor I designate a solution of radical ontic discontnuity between two domains of the world: for example between the sensations that the living being feels and its material basis made of nonsensing entities (particles and forces).[23]

For Meillassoux the divide between worlds (such as between the nonliving and the living or between the physical and the biological) is only explainable in terms of a "higher absurdity." But it would appear that such a metaphor would beg the question of the inexplicable or unexplainable since it itself is a relatively well understood physical phenomena (intense heat on sand being generative of quartz and the like). Yet if Meillassoux's point is simply to argue that the generation of fossils across various fields of explanation can happen unpredictably and in an instant, this we can grant. But to push this degree of surprise to metaphysicalize Hume's problem (to let the corrosive tonic of skepticism sink to the very ground of all existence, would still seem to beg the question).

The issue is of course the gap (and strangely this gap may help prove Meillassoux's point) between the act of writing from the mind to materiality and the fulgurite of the lightning strike. It i s a question of what thought can do and what it means that something happened. This is why science fiction for Meillassoux fundamentally obeys the laws of physics though it may suggest new discoveries about them while his notion of "Extro-Science Fiction" portrays worlds in which the laws of science may break.[24]

But how far is this exactly from Deleuze's own metaphorical use of Mallarme's throw of the dice to index a virtual capacity for the radically new, for the intrusion of the virtual into the world?[25] Does the lawlessness of the world ultimately serve to justify the purely vectorial subject? This is the great gift of Mallarme's "Throw of the Dice" for Meillassoux—the indication that an aspect of modernity that survived and was created out of its fire was the subject as vector, as creative active force, as immortalized in Mallarme's dash (—).[26] It is here again that Deleuze (and especially Deleuze's death-bed Fichteanism) raises his head—whether any question of particularity, of difference, can ever be thought independent of, or in productive tension with, its grasp.

The encapsulating function of consciousness, which still is subject to eventual ruptures (as something that is not a sudden appearance in the recognized order of things but an occurrence that changes the very order of things) or becomings (though the severity or realm of origin consistently raises dispute whether one favors a reading more in league with Manuel DeLanda or one following François Zourabichivilli) stands apart from the punctual inscriber of contingency in Meillassoux's term. But the gap between thinkable materiality in Deleuze or unknowable contingency in Meillassoux seems to rest upon an uncertain level of rational activity in Meillassoux (when a thinker recognizes something as writing rather than as design):

Imagine a young archaeologist, working on an excavation site belonging to a civilization of which, as yet, very little is known, but which it is believed had no writing. Our researcher in the field is working at digging up a tablet; now, when this artefact begins to come to light, she discovers upon it, suddenly, two superposed lines each made of similar marks:

§§§§§§§§

+++++++++,

At first she believes that these are but the similar motifs of the frieze decorating the edge of her tablet. But suddenly, her heart leaps: for she has the intuition that she might have encountered, not a frieze, but two lines of signs. She thinks that she might have found the equivalent of a child's school notebook, in which one learns how to correctly write a character. She now grasps what appeared to be motifs as tokens reproducible at will:

§§§§§§§§§, etc.

+++++++++, etc.

The question, then, is as follows: what happened when her apprehension changed – from the grasping of the marks as motifs, to the grasping of marks as tokens; from seeing a frieze to seeing a double line? From whence, in her mind, came that which totally changes the vision of a civilization yet is not engraved on any tablet: our 'etc.'?[27]

Another option remains in which writing is the result of another unknown capacity which is operative within thought but without any safety net of meaningful sense or human significance. Such a thought is entertained by Claire Colebrook in her chapter entitled "Extinct Theory" in *Death of the PostHuman* (2014). It is a question of inhuman intent or of whether reading material traces counts as reading in the same sense as inferring meaning from reading a poem means meaning. Colebrook contests the notion that intentionality is recognized even if a poem is revealed to have been made by the waves of the ocean.[28]

To return to where we began—for what is in the head, what is meant to go in, what is meant to be inferred, what is meant to be imagined, cannot be sheerly a matter for the guessable architecture of the human mind. But neither can writing be separated from a particularly narrow realm of traces that may be species specific or, even, temporally and historically specific within one species.

This is reversed and made disturbing in Jeff VanderMeer's *Annihilation* (2014) in a scene that mirrors (but also in an inverted way) the death of Cisco's Divinity Student. In exploring the aftermath of an ecological devastation (or rebirth) caused by an alien encounter, a member of an expedition team finds a poem scrawled inside a tower.[29] What is truly horrifying is that it is revealed that one of the area's monsters (the Crawler) wrote the poem by depositing spores as it oozed down the staircase. Part of the question becomes to what degree is the possibility of writing in the material sense bound up with the propensity for intelligence (or how does the given matter of writing that allows it to be effortless yet long-lasting take shape according to the level of intelligence). Do all evolutionary paths from intelligence to writing need to pass through and settle near wells of ink?

How is this recognition different from Meillassoux's surprise at the shift from the assumption of a frieze to the knowing of the fragment of the infinite series? Cisco's novels explore this question but within the inner space of the mind—yet, nonetheless, without ever falling into the psychedelic or the dream world logic where the world of the mind is divided from the world of fulgurite or mushrooms. That is if VanderMeer's *Area X* books suggest that the alien can at least mimic us without being recognizably intelligent, then following Cisco what amount of self-recognition allows for, or is required by, to write about a world that one does not know but could be but is "more than" imagined as merely possible but potential due to the marks of thought.

Notes

1. Jeff VanderMeer, "Introduction," in Michael Cisco, *The Narrator* (Lazy Fascist Press, 2010).
2. Michael Cisco, *The Divinity Student* (Talahassee: BuzzCity Press, 1999), 1–2.
3. This is the general focus of Maurice Blanchot, *The Writing of Disaster* (Lincoln: University of Nebraska Press, 1986).
4. Cisco, *The Divinity Student*, 104.
5. Michael Cisco, *Member* (London: Chomu Press, 2013), 7
6. Ibid., 16.
7. Ibid., 46.
8. Ibid., 44–45.
9. Ibid., 93–94.
10. Ibid., 332.
11. Ibid., 170.
12. Ibid., 142.
13. Ibid., 226.
14. Ibid., 258.
15. Ibid., 47.
16. Michael Cisco, *Unlanguage* (Portland: Eraserhead Press, 2017), 40.
17. Ibid., 31.
18. Ibid., 122.
19. Ibid., 268–69.

20. Quentin Meillassoux, "Iteration, Reiteration, Repetition: A Speculative Analysis of the Sign Devoid of Meaning," in *Genealogies of Speculation: Materialism and Subjectivity Since Structuralism*, eds. Armen Avanessian and Suhail Malik (London: Bloomsbury, 2016), 183.
21. Cisco, *Unlanguage*, 146
22. Had I the space it would be interesting to investigate Cisco's *Animal Money* in relation to Elie Ayache's *The Blank Swan* regarding money, value, and the writing of price.
23. Quentin Meillassoux, "Iteration, Reiteration, Repetition: A Speculative Analysis of the Sign Devoid of Meaning," 149–50.
24. Quentin Meillassoux, *Science Fiction and Extro-Science Fiction* (Minneapolis: Univocal Press, 2015).
25. Ray Brassier, "Stellar Void or Cosmic Animal? Badiou and Deleuze on the Dice-Throw," *Pli: The Warwick Journal of Philosophy* 10 (2000): 200–16.
26. See also Rebecca Comay and Frank Ruda, *The Dash—The Other Side of Absolute Knowing* (Cambridge: MIT Press, 2018).
27. Meillassoux, "Iteration, Reiteration, Repetition: A Speculative Analysis of the Sign Devoid of Meaning," 172.
28. Claire Colebrook, *The Death of the PostHuman: Essays on Extinction, Volume One* (Ann Arbor: Open Humanities Press, 2014).
29. For a connection between the limits of the mind, VanderMeer, and mysticism, see Elvia Wilk, "The World Made Fresh," *E-Flux*, Issue 92, June 2018, accessed May 16, 2019, https://www.e-flux.com/journal/92/205298/the-word-made-fresh-mystical-encounter-and-the-new-weird-divine/.

Bibliography

Blanchot, Maurice. *The Writing of Disaster*. Lincoln: University of Nebraska Press, 1986.

Brassier, Ray. "Stellar Void or Cosmic Animal?" *Pli: The Warwick Journal of Philosophy* 10 (2000): 200–16.

Cisco, Michael. *The Divinity Student*. Tallahassee: BuzzCity Press, 1999.

———. *Member*. London: Chomu Press, 2013.

———. *Unlanguage*. Portland: Eraserhead Press, 2018.

Colebrook, Claire. *The Death of the PostHuman: Essays on Extinction, Volume One*. Ann Arbor: Open Humanities Press, 2014.

Comay, Rebecca, and Frank Ruda. *The Dash—The Other Side of Absolute Knowing*. Cambridge: MIT Press, 2018.

Meillassoux, Quentin. *Science Fiction and Extro-Science Fiction*. Minneapolis: Univocal Press, 2015.

———. "Iteration, Reiteration, Repetition: A Speculative Analysis of the Sign Devoid of Meaning." In *Genealogies of Speculation: Materialism and Subjectivity Since Structuralism*, edited by Armen Avanessian and Suhail Malik, 117–98. London: Bloomsbury, 2016.

VanderMeer, Jeff. "Introduction." In Michael Cisco, *The Narrator*. Lazy Fascist Press, 2010.

Wilk, Elvia. "The World Made Fresh." *E-Flux*, Issue 92, June 2018. Available online at https://www.e-flux.com/journal/92/205298/the-word-made-fresh-mystical-encounter-and-the-new-weird-divine/.

CHAPTER 11

Notes on the Alluring Weirdness of (Materialist) Rumination and Regurgitation: Reading Ariana Reines and Jamie Stewart

Marius Henderson

In perhaps the most prominent recent theorizations of the weird, as espoused by Mark Fisher and Graham Harman, the weird i s dscussed as an ontological as well as an epistemological category.[1] Concomitantly, Fisher and Harman associate particular affective states, such as shock and allure, as well as particular aesthetic modes with the weird, for instance, modernist montage and cubist perspectivation.[2] Whereas this chapter is more indebted to Fisher's conceptualization of the weird as a juxtaposition of incongruent phenomena, it emerges from an awareness toward and by implicitly taking into consideration the partial situatedness of Fisher's and Harman's approaches in relation to different traditions of materialism in critical theory, that is, with dialectical materialism, in the case of Fisher, and

M. Henderson (✉)
Friedrich-Alexander University Erlangen-Nuremberg, Erlangen, Germany

© The Author(s) 2019
J. Greve and F. Zappe (eds.), *Spaces and Fictions of the Weird and the Fantastic*, Geocriticism and Spatial Literary Studies,
https://doi.org/10.1007/978-3-030-28116-8_11

with speculative realism, object-oriented ontology, or "new materialism," in the case of Harman. Yet, both Fisher and Harman attribute the weird with certain (onto-epistemological) relational modes that are somewhat similar. Fisher's connection of the weird with the encroachment of incongruence and Harman's association of the weird with dynamics of fission and fusion have in common that they relate the weird to material-semiotic phenomena and practices, which are constituted by certain relational tensions, for instance, between seemingly incompatible or irreconcilable qualities.

This chapter sets out to explore particular politico-aesthetic modes, rumination, and regurgitation, which transport certain modes of relationality, that not only produce instances of the weird but which also entwine more traditional dialectical materialist concerns of political economy with "new materialist" issues of political ecology. Moreover, these modes imbricate Fisher's and Harman's discussion of the weird in terms of juxtaposition, fusion, and fission, while simultaneously constituting emanations of the weird which lie beyond their conceptualization of the weird. Thus, what stood at the outset of this essay was also the question which artistic and rhetorical strategies tend to generate what comes to be perceived as weirdness. Moreover, the question how these strategies can become translated into particular politico-aesthetic positions and practices is crucial for my inquiry, especially if one takes to heart Fisher's premise that "the weird is that *which does not belong* [emphasis in original]" and that "[t]he weird brings to the familiar something which ordinarily lies beyond it, and which cannot be reconciled with the 'homely' (even as its negation)."[3] As already intimated, Fisher conceives of the weird as a mode for which montage is a particularly fitting form: "The form that is perhaps most appropriate to the weird is montage—the conjoining of *two or more things which do not belong together* [emphasis in original]."[4] Fisher invokes modernist montage techniques, as enacted by surrealists, as examples of renditions of the weird: "Hence the predilection within surrealism for the weird, which understood the unconscious as a montage-machine, a generator of weird juxtapositions."[5] To recapitulate, for Fisher the juxtaposition of incongruent "things" constitutes an integral part of the weird or of processes of weirding.

In this chapter, I would like to invoke rumination and regurgitation as material-semiotic, affectively mediated, aesthetico-political modes which are conducive to the emergence of the weird. I understand/feel rumination and regurgitation primarily as modes of the concomitantly figurative and literal, conceptual and embodied reprocessing of materials, which works across, with, through, and against the plain juxtaposition of incongruents. Additionally, I regard rumination and regurgitation as strategies of rendering what Frank Wilderson III has termed "grammars of suffering," which are produced by specific structural formations of libidinal and political economy.[6] I will base my inquiry into the weirdness-generating potential of rumination and regurgitation on readings of the poetic and other artistic work of Ariana Reines and Jamie Stewart.

"Visceral Abstraction" and the Weird

Before turning to my first case study, Ariana Reines's poetic practice in her volume of poetry entitled *The Cow*, I would like to elaborate on the notion of "visceral abstraction" as it has significantly informed my approach toward Reines's poetry and constitutes a promising concept in discussions of the weird, in particular in relation to questions of political economy and ecology. This notion—that is, visceral abstraction—was developed by Sianne Ngai and she conceptualizes visceral abstraction as a particular mode of catachresis. In my understanding, the term "visceral abstraction" in and of itself already bears a certain potential for weirdness, as it conjoins ostensible antonyms—viscerality and abstraction. Hence, this term transports incongruity and conceptual wrongness, which are attributes of the weird, if one follows Fisher, of course.

Ngai develops her concept of visceral abstraction in the context of a close reading of passages in Marx's *Capital* in which he analyzes the emergence of value out of abstract value-creating labor. The value form and the abstract labor which causes it to emerge, according to Marx, are constitutive structural elements of capitalism. They resemble structural dynamics, which constitute capitalism as an abstract form of social domination. Ngai proclaims that abstract value-creating labor, which arises out of the abstracted equalization of individuals' labor "is difficult to grasp [...] because it is what Marx called 'suprasensible or social' or what we would today call 'emergent.'"[7] Ngai elucidates how Marx employs almost exclusively abstract nouns in his initial renditions of the "specific character of value-creating labour,"[8] which occurs in or as a process of de-specification and reduction of individual labor activities to an abstract general equivalence. Then, however, Marx changes his rhetorical strategy, as Ngai notes, and uses an almost literally materialist imagery, as "the becoming-value of labor" is rendered as an emergent process which "involves something like its transformation from a liquid to a solid state."[9] To illustrate this, Ngai draws attention especially to the following passage from Marx's *Capital*, in which he elaborates on the constitution of value out of abstract labor:

> It is only the expression of equivalence between different sorts of commodities which brings to view the specific character of value-creating labour, by actually reducing the different kinds of labour embedded in the different kinds of commodity to their common quality of being human labour in general. However, it is not enough to express the specific character of the labour which goes to make up the value of the linen [linen serves as an exemplary commodity in this context]. Human labour-power in its fluid state, or human labour, creates value, but is not itself value. It becomes value in its coagulated state, in objective form. The value of the linen as a congealed mass of human labour [*Gallerte* in the German text] can be expressed only as an 'objectivity' [...], a thing which is materially different from the linen itself and yet common to the linen and all other commodities.[10]

Throughout *Capital*, Marx likens value to a "congealed mass of human labour"[11] or to "congealed quantities of homogeneous human labour."[12] The term which Marx uses in the German original in both instances and which Ben Fowkes translated as "congealed mass" and "congealed quantities" is *Gallerte*. According to Ngai, the English translations of *Gallerte* do "a disservice to the visceral impact of Marx's 'bloße Gallerte unterschiedsloser menschlicher Arbeit' [sic] by erasing the specificity of Gallerte: a gelatinous condiment made from the 'meat, bone, [and] connective tissue' of various animals."[13] In this respect, Ngai agrees with Keston Sutherland who has also stressed the viscerality of the concrete noun *Gallerte* and the affective force it might attain in relation to its readers:

> Gallerte is not an abstract noun. Gallerte is now, and was when Marx used it, the name not of a process like freezing or coagulating, but of a specific commodity. Marx's German readers will not only have bought Gallerte, they will have eaten it; and in using the name of this particular commodity to describe not 'homogeneous' but, on the contrary, 'unterschiedslose,' that is, 'undifferentiated' human labour, Marx's intention is not simply to educate his readers but also to disgust them.[14]

The employment of the term *Gallerte* in this specific context holds the potential to cause readers to experience disgust or to be *weirded out* by the abstract processes of abstract social domination which this metaphor refers to. And Ngai conceptualizes Marx's potentially estranging use of the term *Gallerte* in this particular context as an instance of *visceral abstraction*. Moreover, Ngai argues that Marx's use of the term *Gallerte* as a metaphorical expression which helps explaining the dynamic emergence of value out of abstract labor constitutes a form of catachresis—an abusive or broken metaphor, a metaphor which is based on the conjunction of the ostensibly incongruence of viscerality and abstraction. Hence, drawing on Fisher's elaborations on the weird, Marx's catachrestic use of the term *Gallerte* could be regarded as an instantiation of the weird, as a hint at the weird structural dynamics, underlying capitalism, such as the commodification of the products of labor, which lets products that workers produced with their own hands appear as foreign objects, who have ostensibly acquired a life of their own, to them, or, dynamics which conjoin seeming incongruents such as the concrete or visceral and the abstract, elusiveness and objectivity as the following quote, in which Marx uses the term *Gallerte* again, illustrates as well:

> Let us now look at the residue of the products of labour. There is nothing left of them in each case but the same phantom-like objectivity; they are merely congealed quantities [*bloße Gallerte* in the original] of homogeneous human labour, ... of human labour- power expended without regard to the form of its expenditure. All these things now tell us is that human labour-power has been expended to produce them, human labour is accumulated in them. As crystals of this social substance, which is common to them all, they are values—commodity values.[15]

The lexical item *Gallerte* is also weird in its own right, insofar as its etymology remains somewhat obscure and it is attributed with being both fluid and solid as well as static and mobile. Moreover, being a potentially abject substance, which was and perhaps is still primarily handled in the context of usually unpaid "domestic" feminized reproductive labor, *Gallerte* is a notion which can be regarded as kind of metonymic carrier of the traces of said feminized reproductive labor and care work. As materialist feminist critics have pointed out, feminized nonmarket-mediated reproductive labor, such as cooking or childcare, is rendered abject by and concomitantly constitutive for the structural functioning of patriarchal capitalism. Marx's weird catachrestic use of the notion of *Gallerte* could be a faint indicator of the constitutive abjection of feminized reproductive labor—a structural element of capitalism which has largely been neglected by Marxist theorists, as many feminist critics have pointed out. This weird rereading of Marx's use of the term *Gallerte* provides a promising starting point for a critical feminist engagement with Marxian theory which mobilizes the weird and which also informs my approach toward Ariana Reines's poetry.

WEIRD HUMAN–ANIMAL RUMINATIONS IN ARIANA REINES'S *THE COW*

Formally and content-wise Reines's volume of experimental poetry *The Cow* (2006) reprocesses the structural violent objectification, systematic exploitation and commodification of human as well nonhuman bodies, primarily bodies that are rendered as female, under contemporary heteropatriarchal and capitalist conditions, via the predominant figuration of "the cow." The motive of the cow is regurgitated and ruminated on in terms of both its figurative and literal referential capacities.[16]

The Cow juxtaposes seemingly incongruent linguistic material, for instance, entire passages of appropriated text from veterinary manuals on cow diseases or environmentally "safe" animal carcass disposal with appropriated text from such diverse sources as Gertrude Stein, John Ashbery, the King James Bible, the Koran, or Gilles Deleuze and Félix Guattari's *A Thousand Plateaus: Capitalism and Schizophrenia* (1980), and passages of original lyric poetry as well scatological, (post-)pornographic, and deliberately "bad writing."

In a poem in the volume titled "Item," the notion of "the cow" is worked through in terms of its use as a misogynistic slur, which transports vulgar and injurious attributions of "weirdness" to not fully gender conforming bodies, for instance. In addition, in the same poem the term "cow" is explored with respect to its more literal meaning, as the designation of a seemingly "weird" type of animal which is accredited with peculiar corporeal and epistemological capacities: "A cow is a ruminant; she grazes and ambles, she stands still. Is flatulent, lazy, patient. She is named a ruminant not only because her relaxed and melancholy demeanor lend to her a philosophical – if gourmandizing – aspect, but because of the uniqueness of her digestive system."[17]

The poem "Rendered" forms a kind of programmatic centerpiece of the entire volume. Most of the pages of this particular poem are formally structured by the juxtaposition of an upper text section in all caps, which is divided from a lower text section. I read these collisional juxtapositions as poetic montages (also bearing in mind to Sergej Eisenstein's dialectical theory and practice of montage, in particular his intellectual montage of shots of the slaughtering of a cow with the slaughter of striking workers in *Strike*).

The upper blocks of texts, which are printed in all caps, with single phrases being additionally emboldened, consist of appropriated textual material which was taken from veterinary manuals and manuals on animal carcass disposal. The diction of the appropriated text is highly technical and decisively "nonpoetic." It is indicative of an instrumental economic and scientific reason, embedded in patriarchal capitalism, which treats the bodies of "livestock," such as cows—who stand in for all other human and nonhuman beings living under patriarchal capitalism, whether they be alive or dead—solely as a material resource which ought to be subjected to endless transformative processes of value extraction, as the following quote shows, for instance:

LIVESTOCK MORTALITY IS A TREMENDOUS SOURCE OF ORGANIC MATTER. A TYPICAL FRESH CARCASS CONTAINS APPROXIMATELY 32% DRY MATTER, OF WHICH 52% IS PROTEIN, 41% IS FAT, AND 6% IS ASH. RENDERING OFFERS SEVERAL BEN-EFITS TO FOOD ANIMAL AND POULTRY PRODUCTION OPERA-TIONS, INCLUDING PROVIDING A SOURCE OF PROTEIN FOR USE IN ANIMAL FEED, AND PROVIDING A HYGIENIC MEANS OF DISPOSING OF FALLEN AND CONDEMNED ANIMALS. THE END PRODUCTS OF RENDERING HAVE ECONOMIC VALUE AND CAN BE STORED FOR LONG PERIODS OF TIME. USING PROPER PROCESSING CONDITIONS, FINAL PRODUCTS WILL BE FREE OF PATHOGENIC BACTERIA AND UNPLEASANT ODORS.

Across her. Across her.
Upending the weight of her I have to be finding out a grammar.
How to be. Opener.
Sawed open. Easy.
Open like a grammar.
A dust of light the air carries.
What is a night upended.
A carcass in which nothing is left over.
What is a night upended.
Open, a hole where the head disgorges its body.[18]

The lower, non-capitalized text sections appear rather meek in relation to the massive upper text. Some of the lower textual sections look like stanzas of lineated verse whereas others are more akin to blocks of prose poetry. In any case, their diction is rather poetic and they were apparently not appropriated from other sources. One can also trace the presence of a speaker, in the first person singular, who resorts to modes of lyric address, as the quote above indicates. The speaker of the poem proclaims that she strives to "find out a grammar," perhaps a particular "grammar of suffering" which she is subjected to. The poem appears to be invested in understanding particular processes of viscerally abstract domination, however, not by becoming engaged in detached analytical gestures which strive to "saw open" the cow's body in order to excavate some form o f bsolute, hidden truth. *Under*-standing, in the context of *The Cow*'s poetics, can rather be conceived of as a mode of "standing under," as writing from the situated experience of being positioned at the bottom of spatialized relations of power. In a later section, the speaker self-reflexively refers to her mode of poetic speech as a form of under speaking, of speaking from below speaking: "This doesn't have to be speaking. Under speaking. Low. Down. Under speaking."[19] The recurring use of spatial prepositional and adverbial expressions like "under," "down," or also "across" resonates with the spatio-temporal motions which the notions of rumination—as a repetitive movement of back and forth—and regurgitation—as the forceful ejection of something from below—refer to.

As the poem progresses, the strict oppositional structure of the montaged text sections becomes slightly porous and leaky, as traces of an incomprehensible weirdness, trauma-induced errancy, and enigmatic poeticity appear to have encroached upon the self-contained instrumental rationality which nourishes regurgitated passages from a veterinary manual, in which the etiology of the so-called downer cow syndrome is explored:

DOWNER COW: THE TERM IS FREQUENTLY APPLIED TO A MATURE DAIRY COW THAT IS STILL RECUMBENT 3 HR AFTER CALVING DESPITE TREATMENT FOR HYPOCALCEMIA. A SECOND TYPE OF INVOLUNTARY STERNAL RECUMBENCY IS ENCOUNTERED LESS COMMONLY IN CATTLE OF ANY AGE UNDER CONDITIONS NOT ASSOCIATED WITH PARTURITION AND FOR WHICH THE MOST LIKELY ETIOLOGY IS TRAUMA. DOWNER COWS THAT ARE ABLE TO ACTIVELY CRAWL ARE OFTEN REFERRED TO AS 'CREEPERS' AND ARE CONSIDERED

TO HAVE A MORE FAVORABLE PROGNOSIS THAN INACTIVE
ANIMALS. THE CAUSE OF THE RECUMBENCY IS, MORE OFTEN
THAN NOT, ELUSIVE EVEN TO AN EXPERIENCED CLINICIAN.
FURTHERMORE, INEXPERIENCED CLINICIANS MAY MISS AN
OBVIOUS CAUSE IF THEY DO NOT ADOPT A SYSTEMATIC
APPROACH TO DIAGNOSIS;[20]

INVOLUNTARY STERNAL RECUMBENCY MAY BE ASSOCIATED
WITH VERTEBRAL LYMPHOSARCOMA, ABSCESSES, OR *BIZZARE
TRAUMATC INJURIES.*[21]

The first appropriated passage on the "downer cow" is followed by a lower
text section which culminates in the following phrases: "The container
contains the corpse [of the downer cow]. Something gets out from under
the end. Disease. Brains and shit."[22] These phrases gesture toward some
sort of bodily excretion or abject remainder which may elude capture by the
totalizing terror of deadly extractive, yet supposedly environmentally aware,
commodification. The phrases resonate with the following phrase which
appears below a massive appropriated passage that deals with the rectal
exploration of downer cows: "An animal secretes a lot of cortisol if you
harass her too much in killing her and this ruins the meat you are trying to
turn her into."[23] Under the devastating impression of—or rather *under*—
the visually literalized suppression caused by a depressingly lethal totality,
the text traces the remains of liquid leaky corporeal substances which are
imbued with the capacity to sabotage the irrevocable consumption of flesh
(whether it be literal or figurative), and which juxtapositionally, or *weirdly*,
conjoins the cerebral ("Brains") and the excremental ("Shit").

The latter quote in itself contains a reference to a form of postmortem
resistance of the cow, which she may, rather passively and without volition,
enact via a material substance, which she carries in her body: the cortisol.
Hence, "resistance," if I may call it that, in this case, is pictured as an
affective bodily reaction of leaky ruination that sabotages the product which
the patriarchal capitalist machinery wants to turn the cow into.

One could argue that the tracing of excreted resistant materialities in
this volume of poetry appears to provide the ground for felicitous relational
practices of solidarity, also across species boundaries (between women and
cows, for instance), or in Donna Haraway's terms, for emergent configu-
rations of "oddkinship"[24] between humans and nonhuman others. In the
volume, poetry is explicitly designated as a possible medium which could

inaugurate such modes of excremental resistance, as it is quite explicitly likened to feces in another poem.

The Cow's seemingly undecidable, potential weirdness, its wavering between, across, and through the figurative and the literal, the visceral and the abstract, which is enacted in modes of rumination and regurgitation, and its implicit critique of cerebrocentrism, can not only be made productive with respect to the dialectical materialist concept of visceral abstraction. These strategies also become relevant in the context of contemporary new materialist feminism—for instance, with respect to Victoria-Pitts-Taylor's theorization of embodied mindedness and intercorporeal companionship and definitely in relation to Elizabeth A. Wilson's conceptualization of *Gut Feminism* (2015).[25] Wilson critiques the hierarchical opposition of sophistication and a vulgar literal-mindedness, between cuisine and the belly, so to speak, which she perceives to be prevalent in critical theorical discourse. She regards the belly or gut and mind as always already entangled, proposes the notion of a "stomach-mind" and rumination is also an integral practice of *Gut Feminism*, as she states: "the gut is always minded: it ruminates."[26] Moreover, Wilson proclaims:

> If we start with the presumption that mind and gut are keenly alive to each other rather than disengaged, perhaps our political intuitions (for cuisine; against the belly) can be rescripted. In particular, perhaps we can move away from a politics primarily informed by the rhetoric of domination (biology!) and rebellion (culture!) and look for theories that exploit the logic of imbrication.[27]

This notion of gut feminist imbrication resonates with *The Cow*'s poetics of regurgitation and rumination also in terms of the cultivation of a certain pessimism, as Wilson proclaims that: "the feminism that *Gut Feminism* champions offers no plans for repair except through the interpretation of our ongoing, anxious implication in envies, hostilities, and harms."[28] And this tendency toward pessimism is not weird at all but resonates with the following quote from the poem "A Cleaner, Safer World," in *The Cow*, in which the speaker mentions that she would like to become like succor to "the ruined women": "Sometimes I think if I can find a way to really feel my mere going could become as succor to the ruined women I love but it never does. The guilt of knowing the world's evil and still wanting to live in it."[29]

The etymology of "succor," from Latin *sub-* ("from below") and *currere* ("run"); that is, literally meaning: "to run to the help of somebody from below," semantically resonates with the etymology of catachresis, and hence conceptually with Ngai's notion of visceral abstraction and Fisher's notion of the weird, from *kata-* "down, downwards" (expressing the meaning "wrongly") + *khrēsthai* "use"—insofar as both terms encapsulate spatial connotations of vertical movement, moving upwards from below or moving downwards. *The Cow*'s poetics, via resorting to viscerally abstract modes of catachresis thus does not commit to a detached, disembodied, defetishizing, and demystifying analysis, but mobilizes situated or imbricated, critical self-implication, a thinking-alongside, from within and through, in rumination and regurgitation.

Jamie Stewart's Weird Ruminations and Regurgitations—In and Out of the Textual

For more than 15 years Jamie Stewart has been the head of the experimental indie band Xiu Xiu and collaborated with various shifting additional band members. Since 2009 Angela Seo has become the only stable band member besides Stewart. From early on Xiu Xiu's music and lyrics have been designated as being "weird," due to the musical juxtaposition of pop harmonies with avant-garde atonality, partially rather unconventional instrumentation (at least in an indie pop/rock context), and the potentially "disturbing" lyrical content, which, for instance, deals with issues of depression, suicide, sexualized violence, and nonnormative queerness. In addition to his musical output with Xiu Xiu, Stewart has published several collections of poetry, mainly of Haikus, which explore similar textual contents. Due to the fact that Stewart's oeuvre has been associated with weirdness, based on the features mentioned above, for more than a decade now, it begs the question whether weirdness might lose its novelty status and potential shock value at some point, the potential to appear as encroachment of the "outside" onto the "normal" or "homely." Should Stewart's perpetual invocation of weirdness automatically be aligned with a political-economic commodification and exploitation of weirdness? I would argue that, somewhat similar to Reines's poetics in *The Cow*, in the context of Stewart's oeuvre, rumination and regurgitation, as nonlinear, repetitive, figurative and literal, representational and embodied-material, or material-semiotic, practices offer the possibility of an alternative manifestation of the weird, which eschews a dichotomous conceptualization as either subversive or commodified. To repeat, ruminating and regurgitating repetition is key, with respect to Stewart's works, which recursively retains a dynamic tension between the celebration of fetishistic glitter and a seemingly unmediated rawness, which might in itself be considered weird.

Song titles like "Born to Suffer" or "Dear God, I Hate Myself" seem to embalm the fleshy conceptual texture of Stewart's oeuvre with clichéd tropes of teenage angst and self-indulgent deprecation. Affectively his works appear to be structured by a repetitive ruminating attachment to "ugly feelings," like shame, envy, guilt, and self-loathing. The inherent repetitiveness is self-reflexively thematized, for example in the following poem: "PLAY ONE NOTE / SEVENTY-FIVE MINUTES PASS / YET NOTHING HAS CHANGED."[30] Conceptual pathos and postconceptualist bathos tend to collide in Stewart's work, and his texts may appear to some as uncomfortably naïve and blunt (or simply as "bad writing"). In the following haiku: "INFLATE THE BALLOONS / WRITE YOUR NAME ON EACH ONE / PIN, PIN TO POP, POP,"[31] which of course self-reflexively alludes to *pop* music as well, the imperative lyric address transports the inflation and ensuing deflation of an ego. Selfhood starts to twitch and becomes porous, an intensity rather than an entity, interspersed with an intimate (or extimate) weird "outside," which is a notion that the following quotes from the same collection of poems also convey: "PIERCED WITH AN ARROW / FASHIONED FROM THE LEATHERED TAPE WORM / SELFHOOD BOLTS SQUEALING"; "A FAILURE OF WILL / PLEASE REMIND ME AGAIN / RING ON! POINTLESSNESS."[32] Moreover, traversing the separation of the figurative and abstract from the literal and viscerally embodied, Stewart and his collaborators cannibalize their embodied selfhood and subject it to dynamics of reprocessing. Stewart is known for "weird" promo actions and peculiar interactions with fans. He offered a limited edition of a promo bundle, which besides an LP and hand-made chocolate included a t-shirt that has "xiu xiu for life" written on it, in human blood, allegedly Stewart's own. He also asked fans to bring a lock of their hair taped to a card with their name on it to shows and then made a video in which he set each lock on fire, uttered the person's name, and inhaled the smoke ad nauseam.

These recursive, self-exploitative and partially monetarily mediated exchanges of bodily substances apparently are conducive to the emergence of a precarious queer sociality. Moreover, drawing on Judith Butler's more recent work, I would argue that this emergent sociality is not primarily constituted by the sharing of a particular bodily substance, but of corporeal addressability as such, which always already includes "injurability," the capacity to be injured. For Butler, this shared injurability necessarily interlaces with refusal and resistance: "what we see is not simply injury, but an injurability that is actually linked with a form of physical refusal, and what we call resistance is actually this sustained duality of being exposed to injury and, at the same time, refusing and resisting."[33]

Bearing also in mind *The Cow* and Marx's viscerally abstract catachrestic invocation of the *Gallerte*, the wish to become as "succor to the ruined women," one could think of this idea of succor as a kind of prosthetic supplement and of the motions of regurgitation and rumination not as movements which oscillate between clearly delineable positions but as coalescences which imbricate the exposure to structural injury with the resistance to it.

This ruminating and regurgitating coalescence of injury and resistance, which I would regard as one rendition of the weird, is also quite explicitly enacted in the music video to Xiu Xiu's song "Dear God, I Hate Myself."[34] While Jamie Stewart is gleefully eating a chocolate bar, in this video, the band member Angela Seo repeatedly thrusts her fingers into her throat and vomits throughout the entire video, against the backdrop of a stylishly minimalist black and white background and a static camera angle. The video ends with several shots in which Angela Seo vomits onto Stewart's body. I read the video, and especially these final shots, as a critical postpornographic gesture of regurgitation and rumination, as a weird inversion of the so-called "money [or cum]" shot, (that is, in a heteronormative matrix, "male" ejaculation on a "female" body), which has become conventionalized in heteropatriarchal pornographic film, and which, as feminist film theorists like Linda Williams have emphasized, not only perpetuates misogynistic objectification and injury, but also serves as an authenticating "truth sign," and follows a strict instrumental logic of climactic linear progression.[35]

Due to their repetitive temporal structure and their embeddedness in the poppy chiptune melodies of the song and the stylized setting, Seo's regurgitations, which did in fact occur, do not simply act as indexes of a seemingly authentic "realness." And this also holds true for most of the renditions of embodied affective intensities in the works of Xiu Xiu and Stewart in general. Affective intensities are presented as being sweetened, gilded, covered, and caked with other materials, and hence appear to be always already intertwined with processes of (oftentimes culinary) fetishization, such as the outcry "Ooooo," which turns into a candy-coated frame that can be endlessly reproduced, in the song accompanying the video.

What this literal-material and figurative regurgitation and rumination in this video, and Stewart's oeuvre more generally, performatively enacts is a commingling of processes of abstract commodification and visceral corporeality, of political economy and a politicized ecology, which starts by taking into consideration the closest instance of the environment, of one's *oikos*—that is, one's guts, which appear as always already interlaced with the structural dynamics of capitalist (re-)production.

SPECULATIVE OUTLOOK ON THE WEIRD ECOLOGY OF RUMINATION AND REGURGITATION

In a more speculative fashion, drawing on my encounter with the works of Reines and Stewart that I have explored thus far, I would like to end by further thinking with and through the notions rumination and regurgitation as possible renditions of the weird which entwine, twitch, and tweak concerns of political economy and political ecology—of dialectical materialism and "new materialism." Rumination and regurgitation are eminently material-semiotic practices, corporeal phenomena as well as conceptual-representational strategies. Thus, these practices might also be conducive to the imbrication of post-structuralist representationalism with its more recent "new materialist" deflection. Rumination and regurgitation, as it emerges in the works of Reines and Stewart, for instance, are practices which are indicative of a different, or should I say *weird* ecological awareness, which does not reproduce the established ecological discourse of sustainability and preservation, which implicitly presupposes the ideal of a "natural" equilibrium or homeostasis, which ought to be reproduced or restored, in particular in times of a heightened awareness toward the Anthropocene. Contrary to said logic of sustainability and restoration, the weird ecology of rumination and regurgitation does not buy into any romanticization of "nature." The term "buy into" is chosen deliberately, as rumination and regurgitation actively become *invested* into processes of ruination, and emerge from an awareness that the Anthropocene is always already encapsulated in the "Capitalocene,"[36] that "nature," under capitalist conditions, is commodified matter, accumulated, bought, sold, and invested in. And this includes the material-semiotic, natural-cultural bodies of human and non-human animals. The weird ecology of rumination and regurgitation begins always already too close to home, too close to the *oikos*, and to the guts and to the nonlinear spatio-temporal arrangements of abjected feminized reproductive labor. Yet, weirdly ecological rumination and regurgitation also divests from redemptive practices of *recycling* which are often wedded to discourses on sustainability, and which conceptually may invoke the preservation of the status quo or once more, a return to a prior equilibrium. However, for abjected bodies, for racialized and gendered others, for abjected human, animal, and dehumanized populations this equilibrium has never existed, to begin with, is what Reines's and Stewart's works suggest. The instances of ruinous excremental sabotage, the notion of a "succor to the ruined women," the vomit, feces, brains, blood, and guts, which circulate through these works, rather than proposing recycling and sustainability as the restitution of a state of purity and nonpollution, are expressive of a solidarity of bodies which have always already been necropolitically rendered as expendable "waste material." The "gut feminist" practices of regurgitation and rumination seem to follow a weird ecological "inhuman" ethics, which unites bodies that are cast as "garbage" as "outside" the clean, pure, and normative. Insofar, this ethics apparently embodies the following proclamation by Haraway: "We are humus, not Homos, not anthropos; we are compost, not posthuman."[37] Drawing further on Haraway, this ethics calls for a "nonarrogant collaboration with all those in the muddle,"[38] and has seemingly taken the following dictum, which also implicitly imbricates political economy with ecology, by the contemporary Korean poet Kim Hyesoon to heart, or to its guts: "All the Garbage of the World, Unite!"[39]

NOTES

1. See Mark Fisher, *The Weird and the Eerie* (London: Repeater, 2017), 9–10; Graham Harman, *Weird Realism: Lovecraft and Philosophy* (Winchester: Zero Books, 2012), 3–6, 234–40.
2. See Fisher, *The Weird and the Eerie*, 8–17; Harman, *Weird Realism: Lovecraft and Philosophy*, ibid.
3. Fisher, *The Weird and the Eerie*, 10–11.
4. Ibid., 11.
5. Ibid., 11.
6. See Frank B. Wilderson, III., *Red, White & Black: Cinema and the Structure of U.S. Antagonisms* (Durham, NC: Duke University Press, 2010).
7. Sianne Ngai, "Visceral Abstractions," *GLQ: A Journal of Lesbian and Gay Studies* 21, no. 1 (January 2015): 38.

8. Karl Marx, *Capital: Volume 1—A Critique of Political Economy*, trans. Ben Fowkes (London and New York: Penguin Classics, 1992), 142.
9. Ngai, "Visceral," 41.
10. Marx, *Capital*, 142.
11. Marx, *Capital*, ibid.
12. Marx, *Capital*, 128.
13. Ngai, "Visceral," 44.
14. Sutherland [n.p.] quoted in Ngai, "Visceral," 44.
15. Marx, *Capital*, 128; emphasis added.
16. This section, that is, my case study of Reines's *The Cow*, draws on reflections of a chapter in my dissertation, which is entitled "Writing During the Disasters: Modes of Rendering Suffering in Contemporary Experimental Poetry" (University of Hamburg, 2018). The respective chapter analyzes the resonances between Reines's *The Cow*, Ngai's notion of visceral abstraction, and Marx's use of the term *Gallerte* in more detail.
17. Ariana Reines, *The Cow* (New York: Fence Books, 2006), 31.
18. Ibid., 65.
19. Ibid., 71.
20. Ibid., 79.
21. Ibid., 82; emphasis added.
22. Ibid., 80.
23. Ibid., 82.
24. See Donna J. Haraway, *Staying with the Trouble: Making Kin in the Chthulucene* (Durham: Duke University Press, 2016), 2–6.
25. See Victoria Pitts-Taylor, *The Brain's Body: Neuroscience and Corporeal Politics* (Durham: Duke University Press, 2016); Elizabeth A. Wilson, *Gut Feminism* (Durham: Duke University Press, 2015).
26. Wilson, *Gut Feminism*, 22.
27. Ibid., 37–38.
28. Ibid., 179.
29. Reines, *The Cow*, 58.
30. Jamie Stewart, *NIH NIH* (New York: Miniature Garden, 2014), n.p.
31. Ibid.
32. Ibid.
33. Judith Butler and Athena Athanasiou, *Dispossession: The Performative in the Political* (Cambridge: Polity Press, 2013), 111.
34. See "Xiu Xiu, 'Dear God I Hate Myself.'" *Vimeo*, accessed April 6, 2019, https://vimeo.com/54897453.
35. See Linda Williams, *Hard Core: Power, Pleasure, and the "Frenzy of the Visible,"* Expanded Edition (Berkeley: University of California Press, 1999), 100–102.
36. See Haraway, *Staying with the Trouble*, 99–104.
37. Ibid., 55.

38. Ibid., 56.
39. Kim, Hyesoon, *All the Garbage of the World, Unite!*, trans. Don Mee Choi (Notre Dame: Action Books, 2011), 29–30.

Bibliography

Butler, Judith, and Athena Athanasiou. *Dispossession: The Performative in the Political*. Cambridge: Polity, 2013.

Fisher, Mark. *The Weird and the Eerie*. London: Repeater, 2017.

Haraway, Donna J. *Staying with the Trouble: Making Kin in the Chthulucene*. Experimental Futures. Durham: Duke University Press, 2016.

Harman, Graham. *Weird Realism: Lovecraft and Philosophy*. Winchester: Zero Books, 2012.

Kim, Hyesoon. *All the Garbage of the World, Unite!* Translated by Don Mee Choi. Notre Dame: Action Books, 2011.

Marx, Karl. *Capital: Volume 1—A Critique of Political Economy*. Translated by Ben Fowkes. London and New York: Penguin Classics, 1992.

Ngai, Sianne. "Visceral Abstractions". *GLQ: A Journal of Lesbian and Gay Studies* 21, no. 1 (January 2015): 33–63.

Pitts-Taylor, Victoria. *The Brain's Body: Neuroscience and Corporeal Politics*. Durham: Duke University Press, 2016.

Reines, Ariana. *The Cow*. New York: Fence Books, 2006.

Stewart, Jamie. *NIH NIH*. New York: Miniature Garden, 2014.

Wilderson, Frank B. *Red, White & Black: Cinema and the Structure of U.S. Antagonisms*. Durham: Duke University Press, 2010.

Williams, Linda. *Hard Core: Power, Pleasure, and the "Frenzy of the Visible."* Expanded Edition. Berkeley: University of California Press, 1999.

Wilson, Elizabeth A. *Gut Feminism*. Durham: Duke University Press, 2015.

"Xiu Xiu, 'Dear God I Hate Myself.'" *Vimeo*. Accessed April 6, 2019. https://vimeo.com/54897453.

CHAPTER 12

Spaces of Communal Misery: The Weird Post-Capitalism of *Beasts of the Southern Wild*

Marlon Lieber

That which is my essence is my being. The being of the fish is its being in the water, and from this being you cannot separate its essence…. Only in human life does it happen, but even here only in abnormal and unfortunate cases, that being is separated from essence; only here does it happen that a man's essence is not where his being is …. But all beings, excepting cases contrary to nature, are glad to be where and what they are.

—Ludwig Feuerbach, "Principles of the Philosophy of the Future"[1]

A fine panegyric upon the existing state of things! Apart from abnormal cases…, you like to work from your seventh year as a door-keeper in a coal-mine, remaining alone in the dark for fourteen hours a day, and because it is your being therefore it is also your essence.

—Friedrich Engels, "Feuerbach"[2]

In a synopsis of the critical responses to Benh Zeitlin's *Beasts of the Southern Wild*, written five years after its 2012 release, Kyo Maclear remarks that the

M. Lieber (✉)
Christian-Albrechts-Universität, Kiel, Germany

© The Author(s) 2019
J. Greve and F. Zappe (eds.), *Spaces and Fictions of the Weird and the Fantastic*, Geocriticism and Spatial Literary Studies,
https://doi.org/10.1007/978-3-030-28116-8_12

film's reception has been "chasmically divided."[3] A winner of the Caméra d'Or at the Cannes Film Festival as well as the Grand Jury Prize at the Sundance Film Festival, *Beasts* was celebrated by reviewers for its uplifting portrayal of a "joyous" community that is resilient in the face of disaster.[4] Others appreciated its imaginative use of the cinematic form to allow its audience to "visualize climate resistance."[5] However, not all critics shared this enthusiasm. Some noted the strange compatibility of the film's politics with libertarian ideology[6] and its lack of awareness of racial issues. The film's protagonist, a six-year-old girl named Hushpuppy (played by Quvenzhané Wallis, whose impressive performance earned her a nomination for the Academy Award for Best Actress), was said to continue the pernicious tradition of representing black children as "impervious to harm" which has been used to legitimize the denial of care and protection to them.[7] This, Christina Sharpe asserts, was needed to render the film's images "inspiring" rather than "tragic."[8] bell hooks similarly criticizes the film for transforming the "physical and emotional violation" of its child protagonist into a "spectacle,"[9] and Jayna Brown adds that the black family is represented as dysfunctional.[10]

Yet, while Kette Thomas passes too quickly over the repressed legacy of racial domination that provides the context for the film's imagination of blackness, she usefully reminds us that "race as a classification of human beings is entirely absent" in the film's diegetic world.[11] Throughout the film racial difference seems to be entirely irrelevant both for the residents of the Bathtub, the impoverished community off the mainland of Louisiana where the film is set, and for everyone else. Zeitlin himself states in an interview that "all the things that divide people" have intentionally "been removed" from the film, because he wanted the Bathtub to be "this utopian place" that "culturally" is a "total fabrication," despite having a "geographical[]" basis in South Louisiana.[12] But *Beasts* is not just interested in portraying a post-racial community, as it were; its ambition is to emphatically valorize its characters and their community. Perhaps its chief failure, then, is its inability to do so without—unwittingly—reproducing images that risk placing black characters "on a continuum with nonhuman animals"[13]— and, to be sure, the fact that the displacement of racial issues has been a deliberate choice on Zeitlin's part will hardly appease his critics. *Beasts* holds a politically and aesthetically ambiguous position that results from its desire to envision a utopian community and its simultaneous failure to transcend the thought-forms of the existing state of things.

What is needed is a dialectical criticism mindful of Fredric Jameson's advice that the analysis of popular culture must be responsive to both ideological and utopian moments in its objects.[14] *Beasts* articulates a desire for communal autonomy and the refusal of overriding epistemological and political frameworks, which makes its plot readable as a call for an escape from biopolitical and disciplinary forms of governmentality—perhaps even from civilization as such. At the same time, however, the film ends up rejecting the modern state's responsibility to protect particularly vulnerable populations from (un)natural disasters. Moreover, the Bathtub can be understood as an attempt to visualize a community that is not structured by capitalist social forms such as money or wage labor, thus suggesting that alternative criteria for measuring communal wealth are necessary. Yet again it cannot free itself of dominant ideologies by a commitment to qualities historically connoted as masculine. That is to say, the film repeatedly attempts to envision a space both imaginary and real—the Bathtub—in which forms of life, which from the perspectives of humanist philosophy and capitalist political economy must appear as fundamentally wrong—and, hence, following Mark Fisher, as "weird"—can freely develop. It ends up, however, reaffirming dominant political and cultural value systems such as neoliberalism and masculinity.

Cinematic Beasts in a Warming World

Beasts is adapted from the play *Juicy and Delicious* by Lucy Alibar who collaborated with Zeitlin in the production of the film. Its plot establishes a parallel between individual and collective suffering: Hushpuppy's alcoholic and often abusive father Wink suffers from an undisclosed illness that will eventually kill him. Aware of his imminent demise he begins to prepare his daughter for a storm, an unnamed but fairly obvious fictional stand in for Hurricanes Katrina and Rita, that soon floods the Bathtub which is modeled on the Isle de Jean Charles.[15] After Wink and others blow up the levee separating them from what they call the "dry side," they are forcibly evacuated to a hospital. Eventually, they escape the hospital, Wink dies, and Hushpuppy is left with an uncertain future that might see her leave the island.

To tell this story of survival in the face of the (un)natural disasters of a planet shaped by anthropogenic global warming, *Beasts* allows its audience to enter the Bathtub through its use of a camera that closely follows Hushpuppy, showing the "dirty ecology" of the Bathtub whose residents make do with whatever they can "salvage," turning used cars into floats and anything that provides some sort of shelter into a home.[16] According to one critic this produces a "visceral documentary-like effect,"[17] but perhaps it is more apt to call it viscerally subjective. For, in addition to its naturalistic account of the Bathtub's squalor, the film includes scenes that register as products of Hushpuppy's imagination, even though the technique of the "free indirect image" occasionally blurs the distinction between reality and fantasy.[18] Similarly, some critics have argued that the film's magical realism enables it to imaginatively connect "multiple spatial scales," say, the Louisiana coast and Antarctica, thus establishing a perspective well-suited to representing what Andreas Malm calls the "warming condition."[19] Here the global encroaches on the local by way of the Aurochs—prehistoric beasts resurrected when the ice caps begin to melt—which make their way to the Bathtub and encounter Hushpuppy in a climactic scene. The girl, however, approaches the creature and whispers, "You're my friend. Kind of," upon which the Aurochs retreat. Hushpuppy, then, knows to relate to nonhuman animals—and nature, more generally—as something that she and her community need to live with in solidarity rather than as "an object that must be sterilized and tamed."[20] This transformation of the relationship with nature is, in short, what *Beasts* suggests as necessary in the face of climate change.

Yet, Hushpuppy's relationship to nature is not spontaneous. On the contrary, in the Bathtub's school the community's children are taught a "fleshy ontology" that breaks down the categorical distinction between humans and nonhuman animals through an appeal to a shared physical existence[21]: "Meat, meat, meat, meat. Every animal is made out of meat, I'm meat. Y'all asses meat," Miss Bathsheba, their teacher, exclaims, addressing all living creatures—humans included—as animals. Moreover, these different beasts must engage in solidary relations: "That's the most important thing I can ever teach y'all," she tells a group of children after the storm, "Y'all gotta learn to take care of people smaller and sweeter than you are." In many ways, the Bathtub's philosophy resonates with recent philosophical attempts to decenter the figure of the human. If I will not follow this line of investigation any further, however, it is because the film never actually does justice—neither formally nor ideologically—to the ambition of post-humanist theory, beginning with the fact that the human beasts continuously eat other animals. Sometimes their relationship to nature is not so much about mutual aid but a survival of the fittest in the great "buffet of the universe," as Miss Bathsheba calls it.[22]

A Domesticated Weirdness?

The late Mark Fisher has influentially defined "the weird" as that "which does not belong" and, thus, produces a sensation of "*wrongness*" by not conforming to our "concepts and frameworks." In order for *Beasts* to be "weird cinema," then, it would need to render the difference between what is "familiar" and what registers as "wrong" tangible.[23] The film's use of close-up shots and a shaky, hand-held camera that reproduces the vision of a child at the cost of "compositional elements like focus and framing"[24] means that we are invited (or forced) to see through Hushpuppy's eyes and, hence, to identify with her. But, we might begin to ask, should this be so easily possible if the film was really committed to representing a weird onto-epistemology fundamentally at odds with the traditional western mode of relating to the world that most audience members presumably share? Formally the film relies on all too familiar conventions—including the genre of magical realism—that allow for a relatively tension-free identification with Hushpuppy.

In a telling scene that one writer calls the film's "singularly most important" one, Hushpuppy faces off with the Aurochs, but the perspective remains that of the human.[25] We see the creature's eye in a close-up as if looking through Hushpuppy's eyes. However, when the reverse shot cuts to Hushpuppy, we see her in a three-quarter view; in other words, we never assume the Aurochs' perspective on her. After having addressed the animal as a "friend," Hushpuppy continues. "I gotta take care of mine," she says, turns and returns to the other humans while the Aurochs retreat. The lack of interspecies solidarity is, thus, doubled on the level of dialogue; care for one's own trumps the solidarity with the other, and the common "creatureliness" is undermined.[26] The film appears to trouble the categorical split between humanity and nature (including nonhuman animals) through Hushpuppy's precocious voice-over commentary that describes the "hearts" of all beings as engaged in nonverbal communication ideally decodable between different species and by Miss Bathsheba's teachings; ultimately, however, we are only served an always already domesticated version of the weird *qua wrongness* that places few obstacles in the way of integrating—and consuming—it.

NEOLIBERAL BEASTS

Precisely this domesticated version of the weird is what allows for its recuperation in the service of neoliberal and libertarian ideologies. *Pace* Kanye West, the film suggests that the major scandal after Hurricane Katrina was that the federal government cared too much, rather than too little. It is thus transformed into an oppressive apparatus that completely disregards the beasts' desire for autonomy. Before being evacuated, Hushpuppy and Wink have made peace after having gone through another fight the night before. After a vision of the Aurochs—perhaps a dream Hushpuppy is having—the film cuts to a close-up of her sleeping peacefully in her father's arm when the noise of helicopters becomes audible suggesting a threatening intrusion into this rare moment of harmony. And, indeed, Hushpuppy, Wink, and the other beasts are brought to a hospital that, in a curious reversal of the inefficient response to the real 2005 hurricane, seems to work fairly well. In fact, it works with a terrible, even quasi-genocidal efficiency that in the film's logic saves individual lives but threatens to end the collective existence of the beasts (which has the curious result that the film feels like an allegorical depiction of the right-wing conspiracy theory that there exist FEMA concentration camps[27]). While the most of the film is set in the Bathtub where dwellings consisting of waste and natural materials are organically integrated into the landscape of the southern wild, the "film starts to fall apart" when it moves to this neon-lit "civilized bureaucracy."[28] Here, the beasts can no longer move like fish in water, because they exist in a place that is alienated from their essence, as Ludwig Feuerbach might put it; it is not a place that allows subjects to dwell.

In fact, the film's commitment to place rests on what one could call a Heideggerian notion of an essentialist relationship between land and people. In "Building Dwelling Thinking," the German philosopher waxes etymological to get from dwelling to being "preserved from harm and danger," which is to say, being preserved from anything foreign to the dweller's "nature": "To dwell, to be set at peace, means to remain at peace within the free, the preserve, the free sphere that safeguards each thing in its nature."[29] Note how Heidegger here equates being free not with an ability to freely develop but with the imperative to remain the same. It is, thus, only logical to argue that "sav[ing]" an entity is to "set [it] free into its own presencing [*in sein eigenes Wesen freilassen*]."[30] *Beasts*, then, is very much concerned with saving the residents of the Bathtub from the state's attempt to save them by putting them in a hospital; it is, after all, their "nature" that demands that they remain where they are. Thus, the film's woke Heideggerianism—after all, the Bathtub is not the Black Forest, and the beasts are more diverse than what he conceived of as the *Volksgemein-schaft*—proves deeply compatible with a libertarian rejection of the federal government. If an actual survivor of Katrina complained that "[w]e was treated worse than an animal,"[31] in *Beasts* being treated like an animal that is one with its world is an ambition tragically ignored by the state.

Using the terminology of Michel Foucault's 1975–1976 Collège de France lectures, the film's major conflict is organized around a biopolitical attempt to bring "the biological... under state control" insofar as the beasts are subsumed under the rubric of the "population" that can be subjected to a rationalized calculation of health- and illness-related issues.[32] At the same time, it is a disciplinary intervention that treats them as subjects whose behavior must be regulated. In the hospital, Hushpuppy's voice can be heard over the images of an unconscious Bathtub resident connected to a life-support machine—or "plug[ged] into a wall," as she puts it. She relates her father's desire to be put in a boat and set it on fire "if he ever got so old he couldn't drink beer and catch catfish." Strapped to a wheelchair and at the mercy of the doctors, Wink certainly is no longer able to do what he loves and in Zeitlin's film this is a fate worse than death. The biopolitical "management, protection, and cultivation of life"—the conservation of the individual body, that is—becomes "coextensive with the sovereign right to kill" the beasts as a life-form.[33] Biopolitics and what Achille Mbembe has called "necropolitics" are one in the film.[34] In short, *Beasts* mobilizes a Foucauldian critique of modern institutions in an anarchist spirit[35] that is, however, almost indiscernible from libertarian ideology and neoliberal demands to end the state's responsibility for the wellbeing of its citizens.[36]

"RIEN FAIRE, COMME UNE BÊTE"[37]

And yet. Should the defense of the imperfect against that which is worse really suffice? After all, the modern welfare state has been criticized for its devaluation of unpaid reproductive labor traditionally performed by women and its implicit productivism. Its goal was always also to guarantee "an economically productive life" for its subjects who subsequently had to treat their bodily capacities as labor-power to be sold at the market.[38] If the Bathtub despite all its misery figures "as a kind of Eden," it is not just for its pastoral imagery of a happy communion of humans and nature.[39] It is just as much the absence of the compulsion to engage in wage labor that is alluring. In capitalism, human beings are "cut off from nature and from other people," since they "relate to both almost exclusively through the mediation of markets."[40] But in the Bathtub markets and money are absent, and so is wage labor. Wink's love for catching catfish does not ossify into "a particular, exclusive sphere of activity"; that is to say, life in the Bathtub echoes the old Marxian adage that under communism it becomes "possible for me to do one thing today and another tomorrow, to hunt in the morning, fish in the afternoon, rear cattle in the evening, criticise after dinner, just as I have a mind, without ever becoming hunter, fisherman, shepherd, or critic."[41] While the content of the Bathtub's utopia is not the same (less criticizing, more drinking), the form is identical: work "as an activity separate from the rest of life" does not exist.[42] It might be a "communism of penury" (Alberto Toscano), but *Beasts* gives expression to a utopian longing for a collectivity beyond the "community of capital."[43]

And perhaps this is at the same time more powerfully utopian and weirder than anything else in *Beasts*. It suffices to look at Hannah Arendt's discussion of the encounter between European colonizers and African tribes to see how the possibility of a life without labor presents an object that cannot easily be integrated into the modern conceptual framework and registers as a fundamental "*wrongness.*" It is never exactly clear whether her language reflects the "great horror" felt by the Boers or her own revulsion at those who "behaved like a part of nature" and "treated nature as their undisputed master," thus failing to create a "human world." Hence, Arendt's "phantom world of the dark continent" that appears so "unreal and ghostlike"—or weird—to her.[44] One could well follow Mark Fisher and ask whether the encounter with a type of weirdness that troubles thought, producing the feeling that something "should not exist," does not introduce the possibility to look at the familiar from a critical distance, not unlike a Brechtian *Verfremdungseffekt*. In any case, our inherited forms of thought might be what needs to be altered.[45] Rather than seeing the Bathtub as a community whose disdain for labor-intensive world-building simply dooms them to abject poverty, the measure of wealth itself can be revaluated.

Early in the film, Hushpuppy and Wink are floating on the water in the body of an old car. A long shot shows their float that is now being dwarfed by an oil refinery in the distance while the levee separating the two worlds cuts across the image. The film cuts to a shaky point of view shot of the industrial structures while we hear Wink say, "Ain't that ugly over there. We got the prettiest place on earth." Thus the film positions us in opposition to the industrial civilization of the "dry side" at its very beginning. The next sequence opens with an establishing shot of the Bathtub overlaid by Hushpuppy's voiceover commentary. "The Bathtub's got more holidays than the whole rest of the world," she explains, while we are seeing her and others run onto the island in a fast-paced succession of handheld shots that place us right in the middle of the action. "Daddy always saying that up in the dry world they got none of what we got," Hushpuppy continues to images of adults dancing, playing music, and drinking, "They only got holidays once a year." In other words, the beasts measure wealth not in terms of material possessions (whose production might require "ugly" factories) or abstract sums of money. The "dry world" might have all of that, but still remains poor. Wealth for them consists in time of leisure ("holidays") that can be appropriated for the building of communal social relations intensely experienced during collective celebrations—and beautifully visualized by Zeitlin in this sequence diametrically opposed to the sterile hospital later in the film. In short, the capitalist form of social wealth has no meaning in the Bathtub, but "communal luxury" does.[46]

The latter—the intensity of the ties binding the beasts to one another—in many ways only increases when some decide to stay in the Bathtub even though the storm is approaching. In recent years, a number of writers have looked at disaster (both natural and unnatural) as instances during which the familiar social order is suspended and new bonds based on "unbroken solidarities" are created producing a sense of "joyous" communal solidarity that leaves its subjects feeling "enriched rather than impoverished," as Rebecca Solnit puts it. Pointing out that under these conditions the measure of wealth can be transformed, she claims that a "wealth of connections and care" rather than "material wealth" is significant for how communities deal with disaster.[47] The authors of the Out of the Woods collective argue in a similar vein when they write that in addition to the "scarcity" that certainly exists during disastrous events an "abundance of social links" can emerge when everyday life is disrupted; in short, a "collective abundance" that might be appropriated by "disaster communism."[48] The anonymous insurrectionist that form the Invisible Committee explicitly refer to the aftermath of Katrina, praising those who, like the beasts in Zeitlin's film, "refused to leave the terrain." They, too, believe that the "decomposition of all social forms" might open a space "for a wild, massive experimentation with new arrangements."[49] Shared by these examples and *Beasts* is a sense that exceptional situations might reveal a potentiality for solidary social relations that normally remains hidden by the capitalist mediation of human relations by relations between things.[50]

SHE'S THE MAN, OR, BEASTLY MASCULINITY

Beasts, thus, not only reproduces neoliberal ideologemes, but can be read as a fulfillment of the utopian desire for communal solidarity beyond capital and state—even though it can only do so by way of imagining a devastating catastrophe. ("It's easier," of course, "to imagine the end of the world than to imagine that you don't know the rest of the quote," as McKenzie Wark has quipped.[51]) Yet, the film's vision of "communal luxury" remains strangely ambiguous. There is an abundance of noncapitalist wealth in the form of free time at the collective's disposal, to be sure; but at the same time social relationships are strangely impoverished through the emphasis placed on masculinity. More precisely, the ambition in *Beasts* is for Hushpuppy not just to become an "animal," but also a "man" regardless of her actual body.

At school, when Hushpuppy and the other children first learn about the prehistoric Aurochs, Miss Bathsheba tells them that the creatures "would gobble them cave babies down." The "cavemans," she goes on, "was sitting around crying like a bunch of pussies." She warns her pupils that "any day now, fabric of the universe is coming unraveled" so that "y'all better learn how to survive now"—the implication being that this requires them not to be feminine. To turn his daughter into a "man" before he can no longer take care of her is also Wink's ambition, and throughout the film he praises her for tough behavior by telling her that she is "the man." On their final evening before the evacuation, he offers her a drink, arm wrestles her, and when she defeats him yells, "You the man," until she concurs. "I'm the man," the girl screams defiantly.[52] Significantly, in the hospital where the beasts are threatened with—figurative—extinction, Hushpuppy's appearance is radically changed. She wears a prim blue dress and her hair is braided; that is, she has been made to look feminine. In *Beasts*, it is as if the birth of the girl must be ransomed by the death of the animal/man.

Somewhat ironically, then, the film ends up reproducing the commitment to "historically specific, male-connoted ideas of autonomy and freedom" that characterized the modern welfare state whose disciplinary grasp the beasts were precisely trying to escape.[53] But then, this glorification of a virile masculinity in a space deserted by the institutions of modern society yet again aligns the film with others of questionable political leanings. Phil

Neel shows in *Hinterland* that contemporary far-right movements borrow "the language, tactics, and aesthetics of the radical left" in a bid to establish "autonomous zones" in areas where the federal government is weak. But the tribalism of certain groups is necessarily accompanied by the apotheosis of self-reliant masculinity. In the words of Wolves of Vinland member Jack Donovan they are "becoming barbarians. They're leaving behind attachments to the state, to ... this grotesque modern world of synthetic beauty and dead gods. They're building an autonomous zone, a community defined by face-to-face and fist-to-face connections where manliness and honor matter again."[54] Replace the ambition to become a "barbarian" with the desire to become a "beast," and it almost reads like a summary of *Beasts*.[55] Outside of civilization, men can be men again; and Zeitlin's film merely adds that, out there, girls can be men, too.

With Bini Adamczak one could argue that the film embodies an "androcentric universalism"—which, according to her, also characterized the Russian Revolution—insofar as women can be "integrate[d]" but "femininity" cannot so that those "who transcended the female gender assigned to them" are idealized. She grasps the "wealth of gender" as encompassing a plurality of "modes of existence" that can be appropriated selectively. Hence, the privileging of masculinity amounts to a reduction of "the wealth of social relationships" by making inaccessible all modes of relationship declared to be feminine.[56] At least when gender relations are concerned, then, life in the Bathtub amounts to a communal misery. Again, what might appear weird at first glance—the film's envisioning of social relations rich in a way that cannot easily be grasped by the concepts provided by capitalist modernity—remains haunted by the familiar—by a way of life organized around masculine values, in this case. The weirdest (in the colloquial sense of the term) thing about *Beasts of the Southern Wild* ultimately turns out to be its desire to imagine utopia and its complete inability to do so in manner that would transmit the present world and its (neoliberal, libertarian, and right-wing) ideologies.

NOTES

1. Ludwig Feuerbach, "Principles of the Philosophy of the Future," in *The Fiery Brook: Selected Writings*, trans. Zawar Hanfi (London and New York: Verso, 2012), §27; emphasis in original.

2. Frederick Engels, "Feuerbach," trans. S. Ryazanskaya, in Karl Marx and Frederick Engels, *Collected Works*, vol. 5 (London: Lawrence & Wishart, 1975), 13.

3. Kyo Maclear, "Something So Broken: Black Care in the Wake of *Beasts of the Southern Wild*," *ISLE: Interdisciplinary Studies in Literature and Environment* 25, no. 3 (Summer 2018): 607.

4. See, for instance, David Denby, "'Beasts of the Southern Wild,'" *The New Yorker*, June 29, 2012, https://www.newyorker.com/culture/culture-desk/beasts-of-the-southern-wild.

5. Nicholas Mirzoeff, "Becoming Wild," *Occupy 2012*, September 30, 2012, http://www.nicholasmirzoeff.com/O2012/2012/09/30/becoming-wild/.

6. Kelly Candaele, "The Problematic Political Message of 'Beasts of the Southern Wild,'" *Los Angeles Review of Books*, August 9, 2012, https://lareviewofbooks.org/article/the-problematic-political-messages-of-beasts-of-the-southern-wild/.

7. Maclear, "Something So Broken," 615.

8. Christina Sharpe, "Beasts of the Southern Wild—The Romance of Precarity I," *Social Text*, September 27, 2013, https://socialtextjournal.org/beasts-of-the-southern-wild-the-romance-of-precarity-i/.

9. bell hooks, "No Love in the Wild," *NewBlackMan (in Exile)*, September 6, 2012, https://www.newblackmaninexile.net/2012/09/bell-hooks-no-love-in-wild.html.

10. Jayna Brown, "Beasts of the Southern Wild—The Romance of Precarity II," *Social Text*, September 27, 2013, https://socialtextjournal.org/beasts-of-the-southern-wild-the-romance-of-precarity-ii/.

11. Kette Thomas, "With an Eye on a Set of New Eyes: Beasts of the Southern Wild," *Journal of Religion & Film* 17, no. 2 (October 2013): n.p.

12. Jeremy Butman, "'Beasts of the Southern Wild' Director: Louisiana Is a Dangerous Utopia," *The Atlantic*, June 27, 2012, https://www.theatlantic.com/entertainment/archive/2012/06/beasts-of-the-southern-wild-director-louisiana-is-a-dangerous-utopia/259009/.

13. Tavia Nyong'o, "Little Monsters: Race, Sovereignty, and Queer Inhumanism in *Beasts of the Southern Wild*," *GLQ: A Journal of Lesbian and Gay Studies* 21, no. 2–3 (June 2015): 251.

14. Fredric Jameson, "Reification and Utopia in Mass Culture," in *Signatures of the Visible* (New York and London: Routledge, 1992), 29.

15. In fact this island is home to the Biloxi-Chitimacha-Choctaw tribes, but in *Beasts* no natives appear in the Bathtub, which has prompted Tavia Nyong'o to argue that the film pushes "indigeneity … off the map" in a quasi-colonial gesture ("Little Monsters," 264).

16. Patricia Yaeger, "*Beasts of the Southern Wild* and Dirty Ecology," *Southern Spaces*, February 13, 2013, https://southernspaces.org/2013/beasts-southern-wild-and-dirty-ecology.
17. Jake T. Bart, "Once Upon a Time in Louisiana: The Complex Ideology of Beasts of the Southern Wild," *Cinesthesia* 3, no. 2 (2014): n.p.
18. Nyong'o, "Little Monsters," 256.
19. Ali Brox, "The Monster of Representation: Climate Change and Magical Realism in 'Beasts of the Southern Wild,'" *The Journal of the Midwest Modern Language Association* 49, no. 1 (Spring 2016): 145; Andreas Malm, *The Progress of This Storm: Nature and Society in a Warming World* (London and New York: Verso, 2018).
20. Thomas, "With an Eye."
21. Christopher Lloyd, "Creaturely, Throwaway Life After Katrina: *Salvage the Bones* and *Beasts of the Southern Wild*," *South: A Scholarly Journal* 48, no. 2 (Spring 2016): 257.
22. Kristin Ross notes that Peter Kropotkin, the anarchist communist who studied relations of "mutual aid," emphasized "the role played in evolutionary survival by forms of cooperation among the species" precisely as an alternative to Darwinian "competitive struggle" (*Communal Luxury: The Political Imaginary of the Paris Commune* [London and New York: Verso, 2015], 72).
23. Mark Fisher, *The Weird and the Eerie* (London: Repeater Books, 2016), 13; emphasis in original.
24. Bart, "Once Upon a Time."
25. Thomas, "With an Eye."
26. Lloyd, "Creaturely," 248.
27. See Larry Keller, "Fear of FEMA," *Southern Poverty Law Center*, November 18, 2010, https://www.splcenter.org/fighting-hate/intelligence-report/2010/fear-fema.
28. Yaeger, "*Beasts.*"
29. Martin Heidegger, "Building Dwelling Thinking," in *Poetry, Language, Thought*, trans. Albert Hofstadter (New York: Harper Perennial, 2001), 147.
30. Heidegger, "Building," 148.
31. Quoted in Veronica Barnsley, "The Postcolonial Child in Benh Zeitlin's *Beasts of the Southern Wild*," *The Journal of Commonwealth Literature* 51, no. 2 (2016): 249. Moreover, many remained in New Orleans not out of "a nebulous idea of filiality toward the landscape," but simply because they could not afford to leave (Daniel Spoth, "Slow Violence and the (Post)Southern Disaster Narrative in Hurston, Faulkner, and *Beasts of the Southern Wild*," *The Mississippi Quarterly* 68, nos. 1–2 [Winter-Spring 2015]: 155).
32. Michel Foucault, *"Society Must Be Defended": Lectures at the Collège de France, 1975–76*, trans. David Macey (New York: Picador, 2003), 240, 246.

33. Achille Mbembe, "Necropolitics," trans. Libby Meintjes, *Public Culture* 15, no. 1 (Winter 2003): 17.

34. Mbembe defines "necropolitics" as the "subjugation of life to the power of death" and argues that "the capacity to dictate who may live and who must die" is an index of sovereignty. Drawing on Foucault's account of biopolitics, Mbembe turns to the power to "expose to death"—whether literal death or forms of "social death" experienced under conditions of slavery and colonialism ("Necropolitics," 39, 11, 12).

35. See, for instance, Luc Laporte-Rainville, "En pays anarchiste," *Ciné-Bulles* 31, no. 2 (Spring 2013): n.p.

36. On Foucault's "thinly veiled sympathy" for neoliberalism, see Daniel Zamora, "Foucault, the Excluded, and the Neoliberal Erosion of the State," in *Foucault and Neoliberalism*, eds. Daniel Zamora and Michael C. Behrent (Cambridge and Malden, MA: Polity Press, 2016).

37. The phrase is Theodor Adorno's, who thus articulates animality and a post-work imaginary: "*Rien faire, comme un bête*, lying on the water and looking peacefully at the sky"; needless to say, Adorno would be horrified by the utopia imagined in *Beasts* (*Minima Moralia: Reflections from Damaged Life*, trans. E.F.N. Jephcott [London and New York: Verso, 2005], 157).

38. Isabell Lorey, *State of Insecurity: Government of the Precarious*, trans. Aileen Derieg (London and New York: Verso, 2015), 36, 27.

39. Miriam Strube, "Recycling and Surviving in *Beasts of the Southern Wild*: Screening Katrina as a Magic Realist Tale," in *After the Storm: The Cultural Politics of Hurricane Katrina*, eds. Simon Dickel and Evangelia Kindinger (Bielefeld: transcript, 2015), 52.

40. Endnotes, "A History of Separation: The Rise and Fall of the Workers' Movement, 1883–1982," *Endnotes* 4 (October 2015): 180.

41. Karl Marx and Frederick Engels, *The German Ideology: Critique of Modern German Philosophy According to Its Representatives Feuerbach, B. Bauer and Stirner, and of German Socialism According to Its Various Prophets*, trans. Clemens Dutt, W. Lough, and C.P. Magill, in Karl Marx and Frederick Engels, *Collected Works*, vol. 5 (London: Lawrence & Wishart, 1975), 47

42. Gilles Dauvé and François Martin, *Eclipse and Re-Emergence of the Communist Movement* (Oakland: PM Press, 2015), 54.

43. Endnotes, "A History," 166.

44. Hannah Arendt, *The Origins of Totalitarianism* (San Diego, New York, and London: Harcourt Brace & Company, 1973), 194, 192.

45. Fisher, *The Weird and the Eerie*, 15.

46. The latter term appears in the "Manifesto of the Artists' Federation of Paris," written during the short-lived Paris Commune in April 1871; see Ross, *Communal Luxury*, 58.

47. Rebecca Solnit, *A Paradise Built in Hell: The Extraordinary Communities That Arise in Disaster* (New York: Penguin Books, 2009), 3, 5, 265.

48. Out of the Woods, "The Uses of Disaster," *commune*, no. 1 (Fall 2018), https://communemag.com/the-uses-of-disaster/.
49. The Invisible Committee, *The Coming Insurrection* (Los Angeles: Semiotext(e), 2009), 83, 42.
50. See Karl Marx, *Capital: A Critique of Political Economy*, vol. 1, trans. Ben Fowkes (London: Penguin Books, 1990), 165.
51. McKenzie Wark, "The Future Later," *Verso blog*, February 13, 2015, https://www.versobooks.com/blogs/1867-the-future-later-mckenzie-wark-takes-over-the-verso-blog. The "quote" Wark alludes to initially appeared in Fredric Jameson, "Future City," *New Left Review*, no. 21 (May/June 2003): 76. A 2009 blog entry provides a useful history of the sentence that was popularized by Slavoj Žižek and also used by Mark Fisher (in *Capitalist Realism: Is There No Alternative?* [Winchester: Zero Books, 2009]); see Qlipoth, "Easier to Imagine the End of the World...," *Qlipoth*, November 11, 2009, https://qlipoth.blogspot.com/2009/11/easier-to-imagine-end-of-world.html.
52. Nyong'o points out that in the play on which the film is based, the same lines are often spoken, yet its protagonist is a boy who is expected to prove his masculinity. Unlike the screen version, the stage character is "clearly *not* the man" and the demand to embody a "virile, patriarchal masculinity" collapses ("Little Monsters," 255; emphasis in original).
53. Lorey, *State of Insecurity*, 29.
54. Phil A. Neel, *Hinterland: America's New Landscape of Class and Conflict* (London: Reaktion Books, 2018), 24; Donovan is quoted on pp. 25–26.
55. See also Cedric Johnson, "Watching the Train Wreck or Looking for the Brake?" *Souls: A Critical Journey of Black Politics, Culture, and Society* 14, nos. 3–4 (July–December 2012): 212.
56. Bini Adamczak, *Beziehungsweise Revolution: 1917, 1968 und kommende* (Berlin: Suhrkamp, 2017), 132, 162, 173; my translations.

BIBLIOGRAPHY

Adamczak, Bini. *Beziehungsweise Revolution: 1917, 1968 und kommende*. Berlin: Suhrkamp, 2017.
Adorno, Theodor. *Minima Moralia: Reflections from Damaged Life*. Translated by E.F.N. Jephcott. London and New York: Verso, 2005.
Arendt, Hannah. *The Origins of Totalitarianism*. San Diego, New York, and London: Harcourt Brace & Company, 1973.
Barnsley, Veronica. "The Postcolonial Child in Benh Zeitlin's *Beasts of the Southern Wild.*" *The Journal of Commonwealth Literature* 51, no. 2 (2016): 240–55.
Bart, Jake T. "Once Upon a Time in Louisiana: The Complex Ideology of Beasts of the Southern Wild." *Cinesthesia* 3, no. 2 (2014): n.p.

Beasts of the Southern Wild. Directed by Benh Zeitlin. Performed by Quvenzhané Wallis, Dwight Henry. Fox Searchlight Pictures, 2012.

Brown, Jayna. "Beasts of the Southern Wild—The Romance of Precarity II." *Social Text*, September 27, 2013. https://socialtextjournal.org/beasts-of-the-southern-wild-the-romance-of-precarity-ii/.

Brox, Ali. "The Monster of Representation: Climate Change and Magical Realism in 'Beasts of the Southern Wild.'" *The Journal of the Midwest Modern Language Association* 49, no. 1 (Spring 2016): 139–55.

Butman, Jeremy. "'Beasts of the Southern Wild' Director: Louisiana is a Dangerous Utopia." *The Atlantic*, June 27, 2012. https://www.theatlantic.com/entertainment/archive/2012/06/beasts-of-the-southern-wild-director-louisiana-is-a-dangerous-utopia/259009/.

Candaele, Kelly. "The Problematic Political Message of 'Beasts of the Southern Wild.'" *Los Angeles Review of Books*, August 9, 2012. https://lareviewofbooks.org/article/the- problematic-political-messages-of-beasts-of-the-southern-wild/.

Dauvé, Gilles, and François Martin. *Eclipse and Re-Emergence of the Communist Movement.* Oakland: PM Press, 2015.

Denby, David. "Beasts of the Southern Wild." *The New Yorker*, June 29, 2012. https://www.newyorker.com/culture/culture-desk/beasts-of-the-southern-wild.

Endnotes. "A History of Separation: The Rise and Fall of the Workers' Movement, 1883–1982." *Endnotes* 4 (October 2015): 70–192.

Engels, Frederick. "Feuerbach." Translated by S. Ryazanskaya. In Karl Marx and Frederick Engels, *Collected Works*, vol. 5, 11–14. London: Lawrence & Wishart, 1975.

Feuerbach, Ludwig. "Principles of the Philosophy of the Future." In *The Fiery Brook: Selected Writings*, translated by Zawar Hanfi, 175–246. London and New York: Verso, 2012.

Fisher, Mark. *Capitalist Realism: Is There No Alternative?* Winchester: Zero Books, 2009.

———. *The Weird and the Eerie.* London: Repeater Books, 2016.

Foucault, Michel. *"Society Must be Defended": Lectures at the Collège de France, 1975–76.* Translated by David Macey. New York: Picador, 2003.

Heidegger, Martin. "Building Dwelling Thinking." In *Poetry, Language, Thought*, translated by Albert Hofstadter, 143–59. New York: Harper Perennial, 2001.

hooks, bell. "No Love in the Wild." *NewBlackMan (in Exile)*, September 6, 2012. https://www.newblackmaninexile.net/2012/09/bell-hooks-no-love-in-wild.html.

The Invisible Committee. *The Coming Insurrection.* Los Angeles: Semiotext(e), 2009.

Jameson, Fredric. "Reification and Utopia in Mass Culture." In *Signatures of the Visible*, 9–34. New York and London: Routledge, 1992.

———. "Future City." *New Left Review*, no. 21 (May/June 2003): 65–79.

Johnson, Cedric. "Watching the Train Wreck or Looking for the Brake?" *Souls: A Critical Journey of Black Politics, Culture, and Society* 14, nos. 3–4 (July–December 2012): 207–26.

Keller, Larry. "Fear of FEMA." *Southern Poverty Law Center*, November 18, 2010. https://www.splcenter.org/fighting-hate/intelligence-report/2010/fear-fema.

Laporte-Rainville, Luc. "En pays anarchiste." *Ciné-Bulles* 31, no. 2 (Spring 2013): n.p.

Lloyd, Christopher. "Creaturely, Throwaway Life After Katrina: *Salvage the Bones* and *Beasts of the Southern Wild*." *South: A Scholarly Journal* 48, no. 2 (Spring 2016): 246–64.

Lorey, Isabell. *State of Insecurity: Government of the Precarious*. Translated by Aileen Derieg. London and New York: Verso, 2015.

Maclear, Kyo. "Something So Broken: Black Care in the Wake of *Beasts of the Southern Wild*." *ISLE: Interdisciplinary Studies in Literature and Environment* 25, no. 3 (Summer 2018): 603–29.

Malm, Andreas. *The Progress of This Storm: Nature and Society in a Warming World*. London and New York: Verso, 2018.

Marx, Karl. *Capital: A Critique of Political Economy*, vol. 1. Translated by Ben Fowkes. London: Penguin Books, 1990.

Marx, Karl, and Frederick Engels. *The German Ideology: Critique of Modern German Philosophy According to Its Representatives Feuerbach, B. Bauer and Stirner, and of German Socialism According to Its Various Prophets*. Translated by Clemens Dutt, W. Lough, and C.P. Magill. In Karl Marx and Frederick Engels, *Collected Works*, vol. 5, 19–539. London: Lawrence & Wishart, 1975.

Mbembe, Achille. "Necropolitics." Translated by Libby Meintjes. *Public Culture* 15, no. 1 (Winter 2003): 11–40.

Mirzoeff, Nicholas. "Becoming Wild." *Occupy 2012*, September 30, 2012. http://www.nicholasmirzoeff.com/O2012/2012/09/30/becoming-wild/.

Neel, Phil A. *Hinterland: America's New Landscape of Class and Conflict*. London: Reaktion Books, 2018.

Nyong'o, Tavia. "Little Monsters: Race, Sovereignty, and Queer Inhumanism in *Beasts of the Southern Wild*." *GLQ: A Journal of Lesbian and Gay Studies* 21, nos. 2–3 (June 2015): 249–72.

Out of the Woods. "The Uses of Disaster." *Commune*, no. 1 (Fall 2018). https://communemag.com/the-uses-of-disaster/.

Qlipoth. "Easier to Imagine the End of the World..." *Qlipoth*, November 11, 2009. https://qlipoth.blogspot.com/2009/11/easier-to-imagine-end-of-world.html.

Ross, Kristin. *Communal Luxury: The Political Imaginary of the Paris Commune*. London and New York: Verso, 2016.

Sharpe, Christina. "Beasts of the Southern Wild—The Romance of Precarity I." *Social Text*, September 27, 2013. https://socialtextjournal.org/beasts-of-the-southern-wild-the-romance-of-precarity-i/.

Solnit, Rebecca. *A Paradise Built in Hell: The Extraordinary Communities That Arise in Disaster*. New York: Penguin Books, 2009.

Spoth, Daniel. "Slow Violence and the (Post)Southern Disaster Narrative in Hurston, Faulkner, and *Beasts of the Southern Wild*." *The Mississippi Quarterly* 68, nos. 1–2 (Winter–Spring 2015): 145–66.

Strube, Miriam. "Recycling and Surviving in *Beasts of the Southern Wild*: Screening Katrina as a Magic Realist Tale." In *After the Storm: The Cultural Politics of Hurricane Katrina*, edited by Simon Dickel and Evangelia Kindinger, 43–58. Bielefeld: transcript, 2015.

Thomas, Kette. "With an Eye on a Set of New Eyes: Beasts of the Southern Wild." *Journal of Religion & Film* 17, no. 2 (October 2013): n.p.

Toscano, Alberto. "Lineaments of the Logistical State." *Viewpoint Magazine*, September 28, 2014. https://www.viewpointmag.com/2014/09/28/lineaments-of-the-logistical-state/.

Wark, McKenzie. "The Future Later." *Verso blog*, February 13, 2015. https://www.versobooks.com/blogs/1867-the-future-later-mckenzie-wark-takes-over-the-verso-blog.

Yaeger, Patricia. "*Beasts of the Southern Wild* and Dirty Ecology." *Southern Spaces*, February 13, 2013. https://southernspaces.org/2013/beasts-southern-wild-and-dirty-ecology.

Zamora, Daniel. "Foucault, the Excluded, and the Neoliberal Erosion of the State." In *Foucault and Neoliberalism*, edited by Daniel Zamora and Michael C. Behrent, 63–84. Cambridge and Malden, MA: Polity Press, 2016.

NOTES ON CONTRIBUTORS

Julius Greve is a Lecturer and Research Associate at the Institute for English and American Studies, University of Oldenburg, Germany. He is the author of *Shreds of Matter: Cormac McCarthy and the Concept of Nature* (Dartmouth College Press, 2018), and of numerous articles on McCarthy, Mark Z. Danielewski, critical theory, and speculative realism. Greve has co-edited *America and the Musical Unconscious* (Atropos, 2015), *Superpositions: Laruelle and the Humanities* (Rowman & Littlefield International, 2017), and "Cormac McCarthy Between Worlds" (2017), a special issue of *EJAS: European Journal of American Studies*. He is currently working on a manuscript on the relation between modern poetics and ventriloquism.

Marius Henderson studied English and American Studies as well as Gender Studies at the University of Hamburg, Germany, and Johns Hopkins University. Currently, he is a Research Associate in American Studies at Friedrich-Alexander University Erlangen-Nuremberg. His publications include the essay collection *Remembering the Holocaust in a Global Age* (co-edited with Julia Lange).

Moritz Ingwersen teaches American Literature and Culture at the Department of Literature, Art, and Media Studies at the University of Konstanz. He holds a Ph.D. in Cultural Studies from Trent University, Ontario, Canada, and has held teaching positions at the University of Cologne and

© The Editor(s) (if applicable) and The Author(s),
under exclusive license to Springer Nature Switzerland AG 2019
J. Greve and F. Zappe (eds.), *Spaces and Fictions of the Weird and the Fantastic*, Geocriticism and Spatial Literary Studies,
https://doi.org/10.1007/978-3-030-28116-8

the University of the Arts Bremen. Interested in literature and science, the environmental humanities, Indigenous literatures, and posthumanisms, his publications include the edited collection *Culture—Theory—Disability: Encounters between Cultural Studies and Disability Studies* (transcript, 2017) and articles on J. G. Ballard, Mark Z. Danielewski, and Neal Stephenson.

Michaela Keck is a Lecturer at the Department of English and American Studies at University of Oldenburg, Germany. She has published two monographs—*Deliberately Out of Bounds: Women's Work on Classical Myth in Nineteenth-Century American Fiction* (Universitätsverlag Winter, 2017) and *Walking in the Wilderness. The Peripatetic Tradition in Nineteenth-Century American Painting and Literature* (Universitätsverlag Winter, 2006)—as well as essays on a variety of topics. Her research foci are nineteenth-century American Studies, ecocriticism, and the intersection of visual culture and literature.

James Kneale is a cultural/historical geographer and an Associate Professor in the Department of Geography at UCL. His publications include essays on H. P. Lovecraft, William Gibson, and literary geography.

Marlon Lieber is an Assistant Professor of American Studies and Cultural and Media Studies at Christian-Albrechts-Universität zu Kiel, Germany. In 2018 he completed his doctoral dissertation (*Reading 'Race' Relationally: Embodied Dispositions and Socialresearcher at Structures in Colson Whitehead's Novels*) at Goethe-University Frankfurt.

Patricia MacCormack is a Professor of Continental Philosophy at Anglia Ruskin University. She is the author of *Cinesexuality* (2008) and *Posthuman Ethics* (2012), the editor of *The Animal Catalyst: Toward Ahuman Theory* (2014) and the co-editor of *Deleuze and the Schizoanalysis of Cinema* (2008), *Deleuze and the Animal* (2017) and *Ecosophical Aesthetics* (2018).

Jolene Mathieson is a Lecturer and research assistant at the University of Hamburg (Germany) where she teaches courses on poetry, aesthetics, and ecology. She is currently finishing a project on the metaphysics of ekphrasis in Western Europe and the US during the eighteenth and nineteenth centuries, and recently published an article in *Poetics Today* on ekphrasis in digital poetry.

Robert T. Tally Jr. is the NEH Distinguishing Teaching Professor of the Humanities and Professor of English at Texas State University, USA. He is the author of several books, including *Topophrenia: Place, Narrative, and the Spatial Imagination* (Indiana University Press, 2019), *Utopia in the Age of Globalization: Space, Representation, and the World-System* (Palgrave Macmillan, 2013) and the editor of several essay collections, including *Teaching Space, Place, and Literature* (Routledge, 2018), *The Routledge Handbook of Literature and Space* (Routledge, 2017).

Eugene Thacker is a Professor of Media Studies at The New School in New York City. Thacker is the author of many books, including *After Life*, the *Horror of Philosophy* series (starting with *In the Dust of This Planet*), *Cosmic Pessimism* and, most recently, *Infinite Resignation*.

Gry Ulstein is a Ph.D. candidate at Ghent University in Belgium where she is a member of the ERC-funded project "Narrating the Mesh" (NARMESH), led by Professor Marco Caracciolo. Ulstein specializes in New Weird fiction and the environmental humanities.

Ben Woodard is a postdoctoral researcher at the Institute for Philosophy and Art Theory (IPK) at Leuphana University in Lüneburg, Germany. His research focuses on the relationship between naturalism and idealism especially during the long nineteenth century. He is the author of *Slime Dynamics: Generation, Mutation, and the Creep of Life* (Zero Books, 2012), *On an Ungrounded Earth* (Punctum, 2013) and *Schelling's Naturalism: Motion, Space and the Volition of Thought* (Edinburgh University Press, 2019).

Florian Zappe is an Assistant Professor of American Studies at the Georg-August-University Göttingen, Germany. He is the author of books on William S. Burroughs (*'Control Machines' und 'Dispositive'—Eine foucaultsche Analyse der Machtstrukturen im Romanwerk von William S. Burroughs zwischen 1959 und 1968*, Peter Lang, 2008) and Kathy Acker (*Das Zwischen schreiben—Transgression und avantgardistisches Erbe bei Kathy Acker*, transcript, 2013), as well as the co-editor of the essay collection *Surveillance|Society|Culture* (with Andrew Gross, Peter Lang, 2019). In addition to that, he has published widely on literary and visual culture. Currently, he is working on a book project on the cultural history of atheism in America.

INDEX

© The Editor(s) (if applicable) and The Author(s),
under exclusive license to Springer Nature Switzerland AG 2019
J. Greve and F. Zappe (eds.), *Spaces and Fictions of the Weird
and the Fantastic*, Geocriticism and Spatial Literary Studies,
https://doi.org/10.1007/978-3-030-28116-8